CLASSICAL
THERMODYNAMICS
AND
QUANTUM STATISTICS

A First Introductory Course

CLASSICAL THERMODYNAMICS
AND
QUANTUM STATISTICS

A First Introductory Course

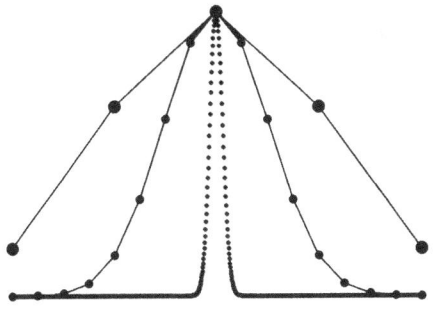

Dmitry Garanin
The City University of New York, USA

World Scientific

NEW JERSEY · LONDON · SINGAPORE · BEIJING · SHANGHAI · HONG KONG · TAIPEI · CHENNAI · TOKYO

Published by

World Scientific Publishing Co. Pte. Ltd.

5 Toh Tuck Link, Singapore 596224

USA office: 27 Warren Street, Suite 401-402, Hackensack, NJ 07601

UK office: 57 Shelton Street, Covent Garden, London WC2H 9HE

Library of Congress Cataloging-in-Publication Data

Names: Garanin, Dmitry author

Title: Classical thermodynamics and quantum statistics : a first introductory course /
    author Dmitry Garanin, the City University of New York, USA.

Description: New Jersey : World Scientific, [2026] | Includes bibliographical references and index.

Identifiers: LCCN 2025054195 | ISBN 9789819823062 hardcover |
    ISBN 9789819824311 paperback | ISBN 9789819823079 ebook for institutions |
    ISBN 9789819823086 ebook for individuals

Subjects: LCSH: Statistical thermodynamics | Quantum statistics | LCGFT: Textbooks

Classification: LCC QC311.5 .G625 2026

LC record available at https://lccn.loc.gov/2025054195

British Library Cataloguing-in-Publication Data

A catalogue record for this book is available from the British Library.

For any available supplementary material, please visit
https://www.worldscientific.com/worldscibooks/10.1142/14580#t=suppl

Desk Editors: Eshak Nabi Akbar Ali/Rhaimie Wahap

Typeset by Stallion Press
Email: enquiries@stallionpress.com

Preface

This is a basic textbook of thermodynamics and statistical physics suitable for a one-semester upper-undergraduate course. This book is based on my lectures read four times since 2009 at Lehman College of the City University of New York. I opted for a conceptually more transparent quantum-mechanical approach to statistical physics, dealing with discrete states from the very beginning. Classical statistics appears as a limiting case of quantum statistics or, in some cases, as an alternative approach at the level of a recipe, to make a comparison with the quantum approach. Elements of quantum mechanics needed are introduced in the text. For the sake of simplicity, only the most important building elements of thermodynamics and statistical physics are included in this small book. Still, this most important material is considered in great detail and even illustrated by numerical solutions for the magnetic systems in the mean-field approximation and for the Bose and Fermi gases.

The course of Statistical Thermodynamics consists of two parts: (1) Thermodynamics and (2) Statistical Physics. Both these branches of physics deal with systems of a large number of particles (atoms, molecules, etc.) at *equilibrium.*

One cubic centimeter of an ideal gas under normal conditions contains $N_L = 2.69 \times 10^{19}$ atoms, the so-called Loschmidt number. Although one may describe the motion of the atoms with the help of Newton's equations, the direct solution of such a large number of differential equations is impossible. On the other hand, one does not need too detailed information about the motion of the individual particles, the *microscopic* behavior of the system. One is rather interested in the *macroscopic* quantities, such as the pressure P. Pressure in gases is due to the bombardment of the walls of the container by the flying atoms of the contained gas. It does not exist if there are only a few gas molecules. Macroscopic quantities such as pressure arise only in systems of a large number of particles. Both thermodynamics and statistical physics study macroscopic quantities and relations between them. Some macroscopic quantities, such as temperature and *entropy*, are nonmechanical.

Equilibrium, or thermodynamic equilibrium, is the state of the system that is achieved some time after time-dependent forces acting on the system have been switched off. One can say that the system approaches the equilibrium if undisturbed. Again, thermodynamic equilibrium arises solely in macroscopic systems. There is no thermodynamic equilibrium in a system of a few particles that are moving according to Newton's law. One should not confuse thermodynamic equilibrium with mechanical equilibrium achieved at a local or global energy minimum. The microstate of a macroscopic system at thermodynamic equilibrium is permanently changing. However, these changes are irrelevant, whereas the important macroscopic state of the system does not change. At equilibrium, the behavior of the system strongly simplifies. Typically, pressure and temperature become the same across the system. Both thermodynamics and statistical physics lead to a number of results at equilibrium, which do not depend on the exact nature of the system and thus are universal. On the contrary, nonequilibrium processes (such as *relaxation*, the process of approaching the equilibrium) are much less universal and therefore much more difficult to describe.

Thermodynamics uses the *phenomenological* approach that consists in working with macroscopic quantities only and not going down to the micro-level of description. Since macroscopic quantities encompass what are directly or indirectly observed in the experiment, one can say that thermodynamics concentrates on the phenomena, leaving out the question of how they are exactly formed. This is the etymology of the word "phenomenological". It turns out that one can formulate a number of principles taken from observations (energy is conserved, temperatures of two bodies in contact tend to equilibrate, etc.), based on which one can arrive at many important conclusions. Thermodynamics emerged in the 19th century in the course of industrialization, and it is the basis of understanding steam engines, refrigerators, and other machines. Thermodynamics studies relations between different macroscopic quantities, taking many inputs from the experiment.

Statistical physics, on the contrary, uses the microscopic approach to calculate macroscopic quantities that thermodynamics has to take from the experiment. The microscopic approach of statistical physics is still much less detailed than the full dynamical description based on Newton's equations. Instead of finding trajectories of the particles, statistical physics operates with the *probability* for the system to be in a certain microscopic state. At equilibrium, the probabilities of microstates turn out to be given by a universal formula, the so-called Gibbs distribution. Using the latter, one can calculate macroscopic quantities.

All thermodynamic relations can be obtained from statistical physics. However, one cannot consider statistical physics as superior to thermodynamics. The point is that thermodynamic relations are universal because they are model-independent. In contrast, results of statistical physics for macroscopic quantities are always based on a particular model and thus are less general.

About the Author

Dmitry Garanin is a Professor in the Department of Physics and Astronomy at Lehman College, City University of New York (CUNY), and a Fellow of the American Physical Society. A theoretical physicist specializing in magnetism, he was born in Moscow in 1954 and graduated from the Moscow Institute of Physics and Technology, U.S.S.R. (1972–1978) with B.S. and M.S. degrees in Physics. He earned his Ph.D. at Moscow State University in 1985, with a thesis on normal modes and relaxation processes in magnetically ordered materials with single-site anisotropy.

After working at the P.N. Lebedev Physical Institute and Moscow Institute of Radioengineering, Electronics, and Automation (MIREA), he immigrated to Germany in 1992. During his European period, he conducted research at the University of Hamburg, Max–Planck Institute for Physics of Complex Systems in Dresden, and the University of Mainz, with additional guest scientist positions in France (Versailles, Perpignan) and the USA (New York).

Since 2005, Professor Garanin has been at Lehman College, where he has also taught classical mechanics at the Graduate Center of CUNY. In 2013, he was elected a Fellow of the American Physical Society "for his theoretical work that shaped research on molecular magnets and helped to develop a deep understanding of their magnetic properties." His most well-known achievement is the development of the Landau–Lifshitz–Bloch equation of motion for ferromagnets, which includes the longitudinal relaxation of the magnetization.

Contents

Chapter 1

Thermodynamics

1.1 Definitions of thermodynamics

1.1.1 System and environment

Thermodynamics studies a macroscopic system that can be in contact with other macroscopic systems and *environments*. Environment (*surrounding*, *bath*, or *heat reservoir*) is a special type of system that has a very large size. The system under consideration can change its state as a result of its contact with the bath, but the state of the bath does not change due to the interaction with a much smaller system. For instance, the thermometer measuring the temperature of a body can be considered as the system, whereas the body itself plays the role of the bath.

1.1.2 Open, close, adiabatic, and isolated systems

Systems can be *open*, *closed*, *adiabatic*, and *isolated*. An open system can exchange mass and energy with the environment.

A closed system cannot exchange mass, but it can receive or lose energy in the form of heat due to the thermal contact with the bath or through the work done on the system.

An adiabatic system is thermally isolated, so that it cannot receive or lose heat, although work can be done on this system. The corresponding processes (such as compression/expansion) are called *adiabatic processes*. Fast processes are also adiabatic because the heat exchange through the surface requires too long a time and becomes inefficient during the time of the process.

An isolated system can exchange neither mass nor energy; there is no contact between the system and the environment.

1.1.3   Extensive, intensive, and specific quantities

Macroscopic physical quantities can be *intensive* and *extensive*. Intensive quantities (pressure P, temperature T) do not depend on the size (mass) of the system, while the extensive quantities (mass M, volume V, energy E or U) scale with the system size. To make this definition more precise, one can split the system into two equal parts by an imaginary membrane. The intensive quantities of the two resulting systems will remain the same, while the extensive quantities of each subsystem will be half of that for the whole system.

Specific quantities are the third type of quantities that can be obtained from an extensive quantity by division by another extensive quantity. By this definition, specific quantities do not depend on the system size. An example of a specific quantity is the mass density

$$\rho = \frac{M}{V}. \tag{1.1.1}$$

1.1.4   Quasistatic, reversible, and irreversible processes

In the Preface, it is stated that thermodynamics studies macroscopic systems at equilibrium, while equilibrium is a time-independent state reached as a result of relaxation under time-independent conditions. On the other hand, thermodynamics considers processes such as compression/expansion or heating that evolve in time. Isn't there a contradiction with the equilibrium nature of thermodynamics? The solution of this paradox is that, "good" thermodynamic processes must be *quasistatic* or *quasiequilibrium*. This means that the external parameters that control the system change with time so slowly that the system at any moment of time is very close to the equilibrium, corresponding to the instantaneous values of the control parameters. These external parameters define the macroscopic quantities of the system at any stage of the process that can be represented as a continuous line in the space of these quantities, such as (P, V). Such space can be called *configurational space*. If the external parameters change fast, the system deviates from the equilibrium and, in most cases, its macroscopic quantities become undefined. In this case, the process cannot be represented as a line in the (P, V) or other diagram. For instance, fast compression of a gas creates sound waves or even shock waves that travel through the system, making density, pressure, and temperature nonuniform, so that there are no unique values of these quantities. In this case, only the initial and final states of the system are equilibrium states, but there is no line connecting them.

Reversible processes are quasistatic processes that can go in both directions along the same line in the space of thermodynamic quantities. Most quasistatic processes are reversible. An exclusion is the slow stirring process of a liquid or gas that cannot be inverted. In this process, the stirring wheel produces work on the system, so that the system warms up. However, from the point of view of the system, this process is equivalent to heating, and it does not matter how exactly the heat is being delivered. By subsequently taking out the heat through the thermal contact with the bath, one can let the system go the same path in the opposite direction in the configurational space, i.e., invert this process. On the other hand, all nonquasistatic processes are *irreversible*.

1.2 Temperature

On the intuitive level, temperature is associated with the notions of "hot" and "cold". The experience shows that if hot and cold bodies are brought into contact, their temperatures would eventually equilibrate. Consider a system in a thermal contact with the bath and make a quasistatic compression or expansion of the system, plotting its states in the (P, V) diagram. As the bath is large, its temperature remains unchanged, and as the process is slow, the temperature of the system will have the same unchanged value. In this way, one obtains the isothermal curve, or *isotherm*, in the (P, V) plot. Repeating this with different temperatures of the bath, one obtains many isotherms. For most substances (except water near 4°C), isotherms corresponding to different temperatures do not cross. This allows us to define the *empirical temperature* T as a parameter labeling isotherms:

$$\phi(P, V) = T. \tag{1.2.1}$$

Indeed, if $T = \text{const}$, P and V are related and belong to a particular isotherm. Note that any monotonic function of the empirical temperature (e.g., $\sqrt{T}$ or T^2) can serve as empirical temperature as well, so that the choice of the latter is not unique. Equation (1.2.1) is the basis of thermometers using different substances, such as alcohol, mercury, and ideal gas. One can, say, fix P to the atmospheric pressure and measure V (or the height of the alcohol or mercury column) that changes with temperature. For the ideal gas, the relation $PV = \text{const.}$ holds, and the constant on the right can be interpreted as the temperature (see the following section). There is a more fundamental approach to the introduction of the temperature in thermodynamics (see M.Sh. Rumer and Yu.B. Ryvkin, *Thermodynamics, Statistical Physics, and Kinetics*) that uses the notions of empirical temperature and empirical entropy together with the isotherms and adiabats of the ideal gas.

Practically, it is convenient to choose the empirical temperature using the experimental fact of thermal expansion and define the change of the temperature as proportional to the change of the volume: $\Delta T \propto \Delta V$. Here, it is important that upon warming or cooling, the volumes of different substances change proportional to each other in a wide range of temperatures to be defined: $\Delta V_1 \propto \Delta V_2 \propto \Delta V_3 \propto \cdots$ Thus, there is no contradiction in the definition of the temperature using different substances. The only thing is to choose the proportionality coefficients in the above formula for different substances and the additive constant (offset) in T. This has been done historically in a number of different ways, resulting in the Fahrenheit, Celsius, Kelvin, and many other defunct temperature scales.

The advantage of the Celsius scale is that we can use very natural events, such as the melting of ice and the boiling of water (at normal conditions), to define the basic temperature points $0°\text{C}$ and $100°\text{C}$. Physicists use the Kelvin scale in which the temperature, corresponding to the extrapolated point, where the volume (or pressure) of the ideal gas vanishes, is set to zero, while $1°$ of temperature difference is the same as in the Celsius scale. The relation between the two scales is

$$T(^{\circ}C) = T(^{\circ}K) - 273.15. \qquad (1.2.2)$$

The existence of temperature as a new (nonmechanical) quantity that equilibrates in the case of systems in thermal contact is called the *zeroth law of thermodynamics*.

1.3 Equation of state

One can rewrite Eq. (1.2.1) in the form symmetric with respect to the thermodynamic quantities P, V, T:

$$f(P, V, T) = 0. \qquad (1.3.1)$$

This relation between the three quantities is called the *equation of state*. If two of the quantities are known, the third can be found from Eq. (1.3.1). If $T = $ const., P and V lie on the curve called *isotherm*. If $P = $ const., V and T lie on the curve called *isobar*. If $V = $ const., P and T lie on the curve called *isochore*.

Equation (1.3.1) is written for a fixed amount of substance in a closed system. As the dependence on the latter is trivial (at fixed P and T, the volume scales with the amount of the substance), this is not included in the equation of state.

Considerations above, including the definition of temperature, pertain to the simplest thermodynamic systems, such as gas or liquid, that are characterized by pressure and volume. There are many other systems described by other macroscopic quantities. For instance, magnetic systems are additionally described by the magnetic induction B (intensive quantity) and the magnetic moment $\mathcal{M}$ (extensive quantity). Usually, magnetic systems are solid and their volume and pressure do not change. Thus, the equation of state for magnetic systems has the form $f(B, \mathcal{M}, T) = 0$.

The simplest equation of state is that of the ideal gas written in the form

$$PV = \nu RT, \qquad (1.3.2)$$

where ν is the number of kilomoles of the gas and $R = 8.314 \times 10^3 \, \text{J}/(\text{kilomole } K)$ is the universal gas constant. The number of kilomoles is defined by $\nu \equiv M/M_{KM}$, where M is the mass of the gas and M_{KM} is the mass of one kilomole of this gas (in kg), numerically equal to its atomic weight (approximately the combined number of protons and neutrons in the nucleus).

Writing the equation of state in the above form contradicts the spirit of thermodynamics that generally neglects the internal structure of matter. One could rewrite the equation of state in the completely thermodynamic form as $PV = M\bar{R}T$, where $\bar{R} \equiv R/M_{KM}$. The downside of this equation is that $\bar{R}$ is not universal and depends on the particular gas. This is why Eq. (1.3.2) is still preferred.

From the point of view of physics, a more useful form of the equation of state of the ideal gas making a connection to the molecular theory is

$$PV = Nk_BT, \qquad (1.3.3)$$

where N is the number of particles (atoms or molecules) in the gas and $k_B = 1.38 \times 10^{-23}$ J/K is the Boltzmann constant. The latter is just the conversion coefficient from kelvins to joules. The combination k_BT is the characteristic thermal energy, corresponding to the temperature T. Note that the unit of PV in the left-hand side of Eq. (1.3.3) is also J. Using the definition of the atomic weight, one can establish the equivalence between the two forms of the equation of state given above.

The isotherm of the ideal gas is a hyperbole, while the isochore and isobar are straight lines.

The temperature in Eqs. (1.3.2) and (1.3.3) is in Kelvin. One can see, as was said above, that at least one of P and V turns to zero at $T = 0$. The beginning of the Kelvin temperature scale has a deep physical meaning: At the absolute zero $T = 0$, the molecules of the ideal gas freeze and stop flying inside the container, falling down into their state of the lowest energy. As the pressure is due to the impact of the molecules onto the walls, it vanishes at $T = 0$. Of course, the equation of state of the ideal gas loses its applicability at temperatures low enough because any actual gas becomes nonideal and condenses at low T. Nevertheless, one can extrapolate the isochore or isobar to the left and find the intercept with the T-axis that is located at $-273°$C, the absolute zero in the Kelvin scale.

1.4 Thermodynamic coefficients

As thermodynamic quantities are related by the equation of state, changing some quantities causes changes in the others. In particular, writing $V = V(P,T)$, one obtains for the infinitesimal changes the full differential

$$dV = \left(\frac{\partial V}{\partial P}\right)_T dP + \left(\frac{\partial V}{\partial T}\right)_P dT, \qquad (1.4.1)$$

where the subscripts T and P mean that these quantities are held constant in the differentiation. The partial derivatives above enter the thermodynamic coefficients, isothermal compressibility $\varkappa_T$ and thermal expansivity β, defined as specific quantities:

$$\varkappa_T = -\frac{1}{V}\left(\frac{\partial V}{\partial P}\right)_T, \quad \beta = \frac{1}{V}\left(\frac{\partial V}{\partial T}\right)_P. \qquad (1.4.2)$$

As $\varkappa_T < 0$ contradicts mechanical stability, all materials have

$$\varkappa_T > 0. \qquad (1.4.3)$$

There is no general principle that could limit the range of β. Most materials expand upon heating, $\beta > 0$. However, materials that consist of long polymer molecules such as rubber contract upon heating, $\beta < 0$, which can be explained by their molecular motion.

Using $P = P(V,T)$, one obtains the differential

$$dP = \left(\frac{\partial P}{\partial V}\right)_T dV + \left(\frac{\partial P}{\partial T}\right)_V dT. \qquad (1.4.4)$$

Both partial derivatives here can be reduced to those considered above with the help of two formulas from calculus. First, one has simply

$$\left(\frac{\partial P}{\partial V}\right)_T = \frac{1}{(\partial V/\partial P)_T}, \qquad (1.4.5)$$

which is an obvious relation. Next, with the help of the triple product rule

$$\left(\frac{\partial P}{\partial T}\right)_V \left(\frac{\partial T}{\partial V}\right)_P \left(\frac{\partial V}{\partial P}\right)_T = -1, \qquad (1.4.6)$$

one obtains

$$\left(\frac{\partial P}{\partial T}\right)_V = -\frac{(\partial V/\partial T)_P}{(\partial V/\partial P)_T} = \frac{\beta}{\varkappa_T}. \tag{1.4.7}$$

This means that one does not need to consider too many thermodynamic coefficients because some can be expressed via those already introduced.

Let us calculate thermodynamic coefficients for an ideal gas, Eq. (1.3.2). Using $V = \nu RT/P$ in Eq. (1.4.2), one obtains

$$\varkappa_T = -\frac{P}{\nu RT}\left(-\frac{\nu RT}{P^2}\right) = \frac{1}{P} \tag{1.4.8}$$

and

$$\beta = \frac{P}{\nu RT}\left(\frac{\nu R}{P}\right) = \frac{1}{T}. \tag{1.4.9}$$

Now, Eq. (1.4.7) yields

$$\left(\frac{\partial P}{\partial T}\right)_V = \frac{P}{T}. \tag{1.4.10}$$

1.5   Work and internal energy

The system and the environment can exchange energy with each other. One of the ways to exchange energy is by doing work that can be understood in mechanical terms. According to Newton's third law, the work done by the environment on the system and the work done by the system on the environment differ in sign. In the general formulation of thermodynamics, we consider the former. For the basic thermodynamic system characterized by P, V, T, the infinitesimal work is given by

$$\delta W = -PdV, \tag{1.5.1}$$

i.e., when the system is compressed, $\delta W > 0$, and the system receives energy. We write δW instead of dW to emphasize that work is not a *state variable*, thus δW is a small increment but not a differential.

A state variable is any thermodynamic quantity that has a well-defined value in any particular state of the system. In particular, for *cyclic processes*, the system returns to the same state at the end of the cycle, so that all state variables assume their initial values. However, nonzero work can be done in cyclic processes (the area circumscribed by the cycle in the P, V diagram), so that one cannot ascribe an "amount of work" to any particular state of the system. The finite work

$$W_{12} = - \int_1^2 PdV \qquad (1.5.2)$$

done on the way from the initial state 1 to the final state 2 depends on the whole way from the initial to the final states. That is, work is a *way function* rather than a state function.

Equation (1.5.1) can be easily obtained considering a cylinder with a moving piston of the area S. Indeed, with the position of the piston l, one obtains $\delta W = -Fdl = -PSdl = -PdV$. In fact, Eq. (1.5.1) is general and it can be obtained for any type of deformation of the system's surface.

Other thermodynamic systems characterized by other macroscopic variables, such as magnetic systems, have their own expressions for the infinitesimal work that we do not consider at the moment.

Let us calculate the work W_{12} for typical processes of the ideal gas. For the isochoric process, obviously, $W_{12} = 0$ because the volume does not change.

For the isobaric process, the integrand P in Eq. (1.5.2) is constant, so that the integral is trivial:

$$W_{12} = -P \int_1^2 dV = P(V_1 - V_2). \qquad (1.5.3)$$

For the isothermal process, with the help of the equation of state (1.3.2), one obtains

$$W_{12} = -\nu RT \int_1^2 \frac{dV}{V} = \nu RT \ln \frac{V_1}{V_2}. \qquad (1.5.4)$$

One can see that positive work is done on the system in both isobaric and isothermal compression.

The work due to the change of externally controlled parameters, such as V, is sometimes called *configurational work*, to distinguish it from the *dissipative work*. Examples of the latter are work done on a liquid by stirring or work on a hard surface by rubbing. It is understood that the dissipative work cannot be described by Eqs. (1.5.1) and (1.5.2). While the configurational work is reversible, dissipative work is irreversible. The total work is the sum of both:

$$\delta W = \delta W_{\text{configurational}} + \delta W_{\text{dissipative}}, \qquad (1.5.5)$$

while

$$\delta W_{\text{dissipative}} \geq 0. \qquad (1.5.6)$$

This nearly evident inequality follows from Eqs. (1.14.8) and (1.14.12).

Similar to mechanics, one can define the *internal energy* U of the system through the work on the way from 1 to 2. To do this, one has to thermally insulate the system from the environment. The experiment shows that the total amount of work (configurational + dissipative) W_{12} on the *adiabatic* system is entirely determined by the initial and final states 1 and 2. This allows one to define the internal energy for any state 2 of the system as

$$U_2 = U_1 + W_{12}, \qquad (1.5.7)$$

where U_1 is an irrelevant additive constant.

To reach state 2 from state 1, one has to make, in general, both configurational and dissipative work. The order in which these works are done is arbitrary, so that there are many paths leading from 1 to 2. Still, the work W_{12} is the same for all these paths, thus U_2, or simply U, is a state quantity. This statement, following from the experiment, can be considered as the foundation of the first law of thermodynamics.

Since the internal energy is a state quantity, one can express it as a function of two basic quantities, say

$$U = U(T, V). \qquad (1.5.8)$$

This is the so-called *caloric equation of state*. Within thermodynamics, the only way to obtain the latter is to take it from the experiment, as described above. Statistical physics provides the caloric equation of state in analytical form in many cases. For ideal gases, the internal energy depends only on the temperature, $U = U(T)$. This is the consequence of the negligibly weak interaction between the particles. For some ideal gases, $U = aT$ with $a = $ const. in a wide range of temperatures. Such an ideal gas is called a *perfect gas*. The constant a turns out to be the heat capacity of the system, as discussed in the following section.

1.6 Heat and the first law of thermodynamics

Having defined the internal energy U for any state P, V of the system, one can relax the condition that the system is adiabatic. As soon as one allows a thermal contact between the system and the environment, it turns out that the energy balance in the mechanical form, $dU = \delta W$, is no longer satisfied. To restore energy conservation, one has to include the heat Q received by the system from the environment. In the infinitesimal form, the energy conservation reads

$$dU = \delta Q + \delta W, \qquad (1.6.1)$$

i.e., the change of the internal energy is work done on the system plus the heat received by the system. Similar to work, heat is not a state function, thus we use δQ instead of dQ. The energy conservation law written in the form of Eq. (1.6.1) constitutes the first law of thermodynamics.

Before the major advances of thermodynamics in the 19th century, the dominant idea of heat was that it was a kind of special substance that could enter or exit the system. If this were true, one could speak of the heat contained in the system, thus heat was a state function. However, this interpretation turned out to be erroneous, as follows from Joule's experiment, proving that heat is just a form of energy. In this experiment, a stirring wheel does dissipative work on a liquid that leads to warming up of the system. The result of this process is identical to that of adding heat through the walls of the container. This is incompatible with the model of heat as a special substance. Moreover, one can show that the total heat received in cyclic processes is usually nonzero, so that heat is not a state function.

Before Joule's experiment, the unit of heat, the calorie, was introduced as the amount of heat needed to increase the temperature of 1 g water by 1°C. Joule's experiment determined the mechanical equivalent of the calory:

$$1 \text{ cal} = 4.19 \text{ J}. \tag{1.6.2}$$

The heat received in a finite process is given by

$$Q_{12} = \int_1^2 (dU - \delta W) = U_2 - U_1 - W_{12}. \tag{1.6.3}$$

Since W_{12} depends on the path between 1 and 2, Q_{12} also depends on this path. Thus, the heat is a path function rather than a state function.

From the point of view of the system, dissipative work on it amounts to adding heat. As the system can then give away this amount of heat (as real heat, not dissipative work), it is convenient to count dissipative work as heat, so that these processes become reversible. In the sequel, dissipative work in most cases will be included in heat.

1.7 Heat capacity

In most cases, adding heat to the system increases its temperature. One can define the heat capacity as

$$C = \frac{\delta Q}{dT}. \tag{1.7.1}$$

C is large if large amounts of heat cause only an insignificant temperature increase. The heat capacity, as defined above, is proportional to the system size and thus is extensive. One can introduce specific quantities, heat and heat capacity per kilomole:

$$q \equiv \frac{Q}{\nu}, \quad c = \frac{C}{\nu} = \frac{\delta q}{dT}. \tag{1.7.2}$$

Heat capacity depends on the type of process. If the heat is added to the system while the volume is kept constant, $dV = 0$, one obtains the isochoric heat capacity:

$$C_V = \left(\frac{\delta Q}{dT}\right)_V. \tag{1.7.3}$$

Also, one can keep a constant pressure, $dP = 0$, to obtain the isobaric heat capacity

$$C_P = \left(\frac{\delta Q}{dT}\right)_P.$$ (1.7.4)

In the isochoric case, no work is done, so the heat fully converts into the internal energy U, and the temperature increases. In the isobaric case, the system usually expands upon heating, and negative work is done on it. This leads to a smaller increase in U and thus a smaller increase in the temperature. Consequently, for most materials, $C_P > C_V$ should be satisfied. Rigorous consideration at the end of this section shows, however, that $C_P > C_V$ is satisfied for all materials, including rubber that shrinks upon heating.

In the isothermal process, the system receives or loses heat, but $dT = 0$, thus $C_T = \pm\infty$.

Finally, in the adiabatic process, $\delta Q = 0$, but the temperature changes. Thus, in this process, $C_S = 0$. The subscript S refers to *entropy*, which is a state function conserved in the reversible adiabatic processes to be studied later.

To consider the most important C_V and C_P, let us rewrite the first law of thermodynamics, Eq. (1.6.1) with Eq. (1.5.1), in the form

$$\delta Q = dU + PdV.$$ (1.7.5)

Considering the energy as a function of T and V, as in Eq. (1.5.8), one can write

$$dU = \left(\frac{\partial U}{\partial T}\right)_V dT + \left(\frac{\partial U}{\partial V}\right)_T dV.$$ (1.7.6)

Combining this with the previous equation, one obtains

$$\delta Q = \left(\frac{\partial U}{\partial T}\right)_V dT + \left[\left(\frac{\partial U}{\partial V}\right)_T + P\right] dV.$$ (1.7.7)

At constant volume, this equation yields $\delta Q = (\partial U/\partial T)_V \, dT$, thus

$$C_V = \left(\frac{\partial U}{\partial T}\right)_V.$$ (1.7.8)

To find the isobaric heat capacity C_P, one has to take into account that at constant pressure, the volume in Eq. (1.7.5) changes because of thermal expansion:

$$dV = \left(\frac{\partial V}{\partial T}\right)_P dT. \tag{1.7.9}$$

Inserting this into Eq. (1.7.7), one obtains

$$\delta Q = C_V dT + \left[\left(\frac{\partial U}{\partial V}\right)_T + P\right]\left(\frac{\partial V}{\partial T}\right)_P dT. \tag{1.7.10}$$

This yields

$$C_P = C_V + \left[\left(\frac{\partial U}{\partial V}\right)_T + P\right]\left(\frac{\partial V}{\partial T}\right)_P. \tag{1.7.11}$$

As said above, the energy of the ideal gas depends only on T, so that $(\partial U/\partial V)_T = 0$. Another partial derivative can be obtained from the equation of state (1.3.2) and is given by $(\partial V/\partial T)_P = \nu R/P$. The result is the famous *Mayer's relation* for the ideal gas:

$$C_P = C_V + \nu R \tag{1.7.12}$$

or $C_P = C_V + R$ for heat capacities per kilomole. In terms of the number of particles N, see Eq. (1.3.3), Mayer's relation becomes

$$C_P = C_V + N k_B \tag{1.7.13}$$

or $C_P = C_V + k_B$ for heat capacities per particle.

We have used $(\partial U/\partial V)_T = 0$ for the ideal gas as taken from the experiment, and we can explain it within molecular theory. It turns out, however, that $(\partial U/\partial V)_T$ can be calculated within thermodynamics and has the form

$$\left(\frac{\partial U}{\partial V}\right)_T = T\left(\frac{\partial P}{\partial T}\right)_V - P. \tag{1.7.14}$$

Now, for the ideal gas from the equation of state follows $(\partial U/\partial V)_T = 0$. In the general case, Eqs. (1.7.11) and (1.7.14) combine to

$$C_P = C_V + T\left(\frac{\partial P}{\partial T}\right)_V \left(\frac{\partial V}{\partial T}\right)_P. \tag{1.7.15}$$

The last term here can be expressed via the thermodynamic coefficients of Section 1.4. Using Eqs. (1.4.2) and (1.4.7), one obtains

$$C_P = C_V + VT\frac{\beta^2}{\varkappa_T}. \tag{1.7.16}$$

Since the compressibility $\varkappa_T > 0$ for all materials, one concludes that always $C_P > C_V$. The derivation of Eq. (1.7.14) is based on the concept of *entropy* discussed later in the course.

1.8 Adiabatic process of the ideal gas

For the ideal gas, the internal energy is a function of the temperature only, thus one has

$$dU = \left(\frac{\partial U}{\partial T}\right)_V dT = C_V dT. \tag{1.8.1}$$

In the adiabatic process, $\delta Q = 0$. Adopting these two results in Eq. (1.6.1) and using Eq. (1.5.1), one obtains

$$C_V dT = -P dV. \tag{1.8.2}$$

Here, one of the quantities P or V can be eliminated with the help of the equation of state (1.3.2). Substituting $P = \nu RT/V$, one obtains

$$C_V dT = -\nu RT\frac{dV}{V}. \tag{1.8.3}$$

This is equivalent to a differential equation that can be integrated if the temperature dependence $C_V(T)$ is known. For the perfect gas, $C_V = \text{const.}$, integrating is trivial and yields

$$\ln T = -\frac{\nu R}{C_V}\ln V + \text{const.} \tag{1.8.4}$$

or, finally, the adiabat equation

$$TV^{\nu R/C_V} = \text{const.} \tag{1.8.5}$$

It is convenient to introduce the ratio of heat capacities:

$$\gamma \equiv \frac{C_P}{C_V}, \tag{1.8.6}$$

so that, with the help of Mayer's equation (1.7.12), one has

$$C_V = \frac{\nu R}{\gamma - 1}, \quad C_P = \frac{\nu R \gamma}{\gamma - 1}. \tag{1.8.7}$$

Then the adiabat equation (1.8.5) becomes

$$TV^{\gamma-1} = \text{const.} \tag{1.8.8}$$

One can also express the adiabat equation via P, V using the equation of state (1.3.2) that yields

$$PV^{\gamma} = \text{const.} \tag{1.8.9}$$

The third form of the adiabat equation is

$$TP^{1/\gamma-1} = \text{const.} \tag{1.8.10}$$

Let us calculate the work done in the adiabatic process from 1 to 2. Using Eq. (1.8.9), one obtains

$$W_{12} = -\int_{V_1}^{V_2} PdV = -\text{const.} \int_{V_1}^{V_2} \frac{dV}{V^{\gamma}} = -\frac{\text{const.}}{1 - \gamma} \left(V_2^{1-\gamma} - V_1^{1-\gamma} \right). \tag{1.8.11}$$

Here, the constant can be eliminated using const. $= P_1 V_1^{\gamma} = P_2 V_2^{\gamma}$ for the two terms. One obtains

$$W_{12} = \frac{1}{\gamma - 1} \left(P_2 V_2 - P_1 V_1 \right). \tag{1.8.12}$$

With the help of the equation of state and Eq. (1.8.7), one can simplify this formula to

$$W_{12} = \frac{\nu R}{\gamma - 1} \left(T_2 - T_1 \right) = C_V \left(T_2 - T_1 \right). \tag{1.8.13}$$

Equation (1.8.13) is a very simple result that can be obtained by another and simpler method. Indeed, according to the first law of thermodynamics, in the adiabatic process, the work is equal to the change of the internal energy, $W_{12} = U_2 - U_1$. For the ideal gas,

$U = U(T)$ and $dU/dT = C_V$. Thus, the internal energy of the ideal gas is given by

$$U(T) = \int C_V(T)dT. \qquad (1.8.14)$$

For the perfect gas, $C_V = \text{const.}$, one has

$$U(T) = C_V T + U_0, \quad U_0 = \text{const.}, \qquad (1.8.15)$$

thus

$$U_2 - U_1 = C_V\left(T_2 - T_1\right), \qquad (1.8.16)$$

so that Eq. (1.8.13) follows.

1.9 Problems

1.9.1 Process $P = AT^b$

A process on an ideal gas is defined by

$$P = AT^b.$$

Express this process in terms of (P, V) and (V, T). Calculate compressibility and thermal expansivity in this process. What is the limitation on b? For which values of b does this process become a known process? Find adiabatic values of the two thermodynamic coefficients above.

Solution: Using the equation of state of the ideal gas

$$PV = \nu RT,$$

one obtains

$$P = A\left(\frac{PV}{\nu R}\right)^b.$$

This can be represented in the simplified form

$$P^{1-1/b}V = \nu R/A^{1/b}.$$

Another form of this process reads

$$\frac{\nu RT}{V} = AT^b,$$

which can be simplified to

$$T^{b-1}V = \nu R/A.$$

Compressibility is given by

$$\kappa = -\frac{1}{V}\frac{dV}{dP}.$$

Using $V \propto P^{-(1-1/b)}$ above, one obtains

$$\kappa = \left(1 - \frac{1}{b}\right)\frac{P^{-(1-1/b)-1}}{V} = \left(1 - \frac{1}{b}\right)\frac{1}{P}.$$

Since mechanical stability requires $\kappa > 0$, the condition on b is $b > 1$ or $b < 0$.

The thermal expansion coefficient is defined by

$$\beta = \frac{1}{V}\frac{dV}{dT}.$$

Using $V \propto T^{-(b-1)}$ above, one obtains

$$\beta = \frac{1-b}{T}.$$

Identification with known processes includes the following: iso-baric process: $b = 0$; isochoric process: $b = 1$; isothermic process: $b \to \infty$; adiabatic process: $b = \gamma/(\gamma - 1) > 1$. From the latter, one obtains adiabatic thermodynamic coefficients

$$\kappa_S = \frac{1}{\gamma P}, \quad \beta_S = -\frac{1}{(\gamma - 1)T},$$

which have to be compared with

$$\kappa_T = \frac{1}{P}, \quad \beta_P = \frac{1}{T}.$$

Note that $\beta_S < 0$ because in the adiabatic process, as the volume decreases, the temperature increases.

1.9.2 Work and heat in the $P = AT^2$ process

A process on an ideal gas is defined by

$$P = AT^2, \quad A = \text{const.}$$

Calculate the received work and heat upon changing the temperature from T_1 to T_2. Assume $C_V = \text{const.}$

Solution: Use the equation of state of the ideal gas

$$PV = \nu RT$$

to express P in terms of V as

$$P = \frac{(\nu R)^2}{AV^2}$$

and integrate

$$W_{12} = -\int_1^2 PdV = -\int_1^2 \frac{(\nu R)^2}{AV^2}dV = \frac{(\nu R)^2}{A}\left(\frac{1}{V_2} - \frac{1}{V_1}\right).$$

Then express V via T,

$$V = \frac{\nu RT}{P} = \frac{\nu RT}{AT^2} = \frac{\nu R}{AT},$$

and substitute it into the work,

$$W_{12} = \nu R(T_2 - T_1).$$

To calculate the heat, use the first law of thermodynamics in the form

$$U_2 - U_1 = Q_{12} + W_{12}.$$

Using

$$U = C_V T + \text{const.}$$

for a perfect gas and the result for the work, one obtains

$$Q_{12} = C_V (T_2 - T_1) - W_{12} = (C_V - \nu R)(T_2 - T_1).$$

1.9.3   Heat capacity in the process $P = AT^b$

Calculate the heat capacity in the process

$$P = AT^b$$

of an ideal gas, expressing it as a function of T. Analyze different cases of b.

Solution: Use the first law of thermodynamics

$$dU = \delta Q - PdV.$$

The infinitesimal received heat is given by

$$\delta Q = dU + PdV = \left(\frac{\partial U}{\partial T}\right)_V dT + \left[\left(\frac{\partial U}{\partial V}\right)_T + P\right] dV.$$

One has

$$\left(\frac{\partial U}{\partial T}\right)_V = C_V,$$

while for the ideal gas, $\left(\frac{\partial U}{\partial V}\right)_T = 0$. Thus, the expression above simplifies to

$$\delta Q = C_V dT + PdV.$$

Now, one has to express dV through dT. From the equation of the process and the equation of state of the ideal gas, one obtains

$$V = \frac{\nu R}{A} T^{1-b}.$$

From here follows

$$dV = \frac{\nu R}{A}(1 - b)T^{-b}dT.$$

With the help of the process equation, this yields

$$\delta Q = C_V dT + \nu R (1 - b) dT$$

and, finally,

$$C = \frac{\delta Q}{dT} = C_V + \nu R (1 - b).$$

From here, for the isobaric process, $b = 0$, one obtains $C = C_P = C_V + \nu R$ (Meyer's relation). For the isochoric process, $b = 1$, one obtains $C = C_V$. For the isothermic process, $b \to \infty$, one obtains $C \to \infty$. For the adiabatic process, $b = \gamma/(\gamma-1)$, where $\gamma = C_P/C_V$, one obtains

$$C = C_V - \frac{\nu R}{\gamma - 1} = \frac{C_V \left(\frac{C_P}{C_V} - 1 \right) - \nu R}{\gamma - 1} = \frac{C_P - C_V - \nu R}{\gamma - 1} = 0,$$

taking into account Meyer's relation above.

1.9.4 Van der Waals gas

Van der Waals equation of state for a nonideal gas describing its transition to liquid has the form

$$\left(P + \frac{a}{V^2} \right) (V - b) = \nu R T,$$

where a describes the attraction of the gas molecules and b describes the volume occupied by the molecules and thus excluded from their motion.

1. Using a plotting program or by hand, plot isotherms of this gas (Fig. 1.1) for different T, setting $a = b = \nu R = 1$. At high T, isotherms are close to those for an ideal gas, but for lower T, they become distorted. Finally, at some $T = T_c$ (critical temperature), the isotherm becomes horizontal at some point called "critical point", where its second derivative also turns to zero.
2. Calculate the isothermal compressibility of the van der Waals gas in terms of (V, T). Obtain its high-temperature limit. What happens with it at the critical point?

3. Find the critical point parameters using the analysis in (1) as a hint.

Solution: Represent the compressibility in the form

$$\kappa_T = -\frac{1}{V}\left(\frac{\partial V}{\partial P}\right)_T = -\frac{1}{V}\left(\frac{\partial P}{\partial V}\right)_T^{-1}$$

and resolve the equation of state for P as

$$P = \frac{\nu RT}{V-b} - \frac{a}{V^2}.$$

One can see that the b-term increases pressure, whereas the a-term decreases pressure, as expected. Differentiating P, one obtains

$$\left(\frac{\partial P}{\partial V}\right)_T = -\frac{\nu RT}{(V-b)^2} + \frac{2a}{V^3},$$

which yields

$$\kappa_T = -\frac{1}{V}\frac{1}{-\frac{\nu RT}{(V-b)^2} + \frac{2a}{V^3}} = \frac{(V-b)^2/V}{\nu RT - 2a\,(V-b)\,/V^3}.$$

At high temperatures, V becomes large, so that one can neglect the terms with a and b and obtain $\kappa_T = V/(\nu RT) = 1/P$, which is the

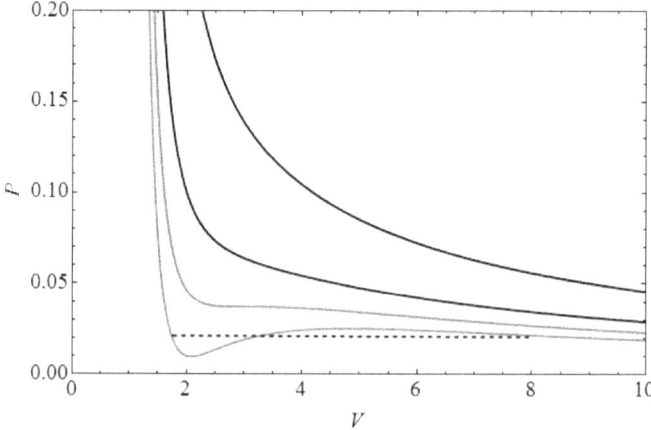

Figure 1.1.   Isotherms of the van der Waals gas.

result for an ideal gas. With lowering T, the volume decreases, and the negative term in the denominator causes κ_T to diverge at the critical point.

To find the critical point, one can use $\left(\frac{\partial P}{\partial V}\right)_T = \left(\frac{\partial^2 P}{\partial V^2}\right)_T = 0$, i.e.,

$$\left(\frac{\partial P}{\partial V}\right)_T = -\frac{\nu RT}{(V-b)^2} + \frac{2a}{V^3} = 0,$$

$$\left(\frac{\partial^2 P}{\partial V^2}\right)_T = \frac{2\nu RT}{(V-b)^3} - \frac{6a}{V^4} = 0.$$

Getting rid of the denominators, one obtains the system of equations

$$\nu RT V^3 = 2a\,(V-b)^2,$$
$$\nu RT V^4 = 3a\,(V-b)^3.$$

Dividing the second equation by the first one yields

$$V = (3/2)(V - b).$$

Solving this equation, one obtains the critical volume

$$V_c = 3b.$$

After that, one obtains the critical temperature

$$\nu RT_c = \frac{2a}{V_c^3}(V_c - b)^2 = \frac{8a}{27b}.$$

Finally, the critical pressure can be obtained from the equation of state,

$$P_c = \frac{\nu RT_c}{V_c - b} - \frac{a}{V_c^2} = \frac{8a}{27b \times 2b} - \frac{a}{(3b)^2} = \frac{a}{27b^2}.$$

1.9.5 Isochore–isotherm cycle

Find the efficiency of a heat machine using an isochore–isotherm cycle of an ideal gas.

Solution: The cycle is similar to the Carnot cycle with the adiabats replaced by isochores, and it is performed in the clockwise direction.

The heat is received by the system at the upward isochore, $Q_2' = Q_{DA}$, and right-bound isotherm, $Q_2'' = Q_{AB}$. The heat is given away on the downward isochore, $Q_1' = -Q_{BC}$, and left-bound isotherm, $Q_1'' = -Q_{CD}$. The efficiency is given by

$$\eta = \frac{Q_2 - Q_1}{Q_2}, \quad Q_1 = Q_1' + Q_1'', \quad Q_2 = Q_2' + Q_2''.$$

The heat on isochores can be calculated via the change of the internal energy because there is no work:

$$Q_2' = Q_{DA} = C_V (T_2 - T_1), \quad Q_1' = -Q_{BC} = C_V (T_2 - T_1).$$

The heat on the isotherms can be calculated via the work done because the internal energy does not change:

$$Q_2'' = Q_{AB} = -W_{AB} = -\nu R T_2 \ln \frac{V_A}{V_B} = \nu R T_2 \ln \frac{V_B}{V_A},$$

$$Q_1'' = -Q_{CD} = W_{CD} = \nu R T_1 \ln \frac{V_C}{V_D} = \nu R T_1 \ln \frac{V_B}{V_A}.$$

Now, one obtains

$$\eta = \frac{\nu R T_2 \ln \frac{V_B}{V_A} + C_V (T_2 - T_1) - \nu R T_1 \ln \frac{V_B}{V_A} - C_V (T_2 - T_1)}{\nu R T_2 \ln \frac{V_B}{V_A} + C_V (T_2 - T_1)},$$

which simplifies to

$$\eta = \frac{T_2 - T_1}{T_2 + C_V (T_2 - T_1) / \left(\nu R \ln \frac{V_B}{V_A} \right)}.$$

One can see that because of the additional positive term on the denominator, the efficiency is smaller than the efficiency of the Carnot cycle, $\eta = (T_2 - T_1) / T_2$.

1.10   Heat machines

Heat machine was a major application of thermodynamics in the 19th century that greatly contributed to its development. As engines convert energy between different forms, such as work and heat, the basis of their understanding is the first law of thermodynamics, Eq. (1.6.1). There are three types of heat machines: engines, refrigerators, and heat pumps. In all cases, a heat machine includes two reservoirs with different temperatures and a system that exchanges heat with the two reservoirs and does work in a cyclic process. Engines of the 19th century used steam as a source of heat, while the steam was obtained by heating water by fire. Contemporary motors use fuel that burns and generates heat directly. Refrigerators and heat pumps also use agents other than water.

1.10.1   Heat engine

In engines, during one cycle, the system receives the heat Q_2 from the hot reservoir, gives the heat Q_1 to the cold reservoir, and makes the work, see Fig. 1.2. This work is $-W$, the negative of the work W made on the system that we are using in general formulations. The efficiency of the engine is defined as the ratio of the output energy (work $-W$) to the input energy Q_2:

$$\eta = \frac{-W}{Q_2}, \qquad (1.10.1)$$

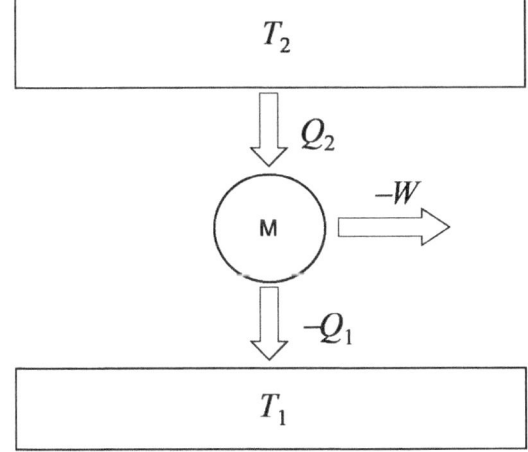

Figure 1.2.   Schematic representation of a heat engine.

whereas $-Q_1$ is the lost energy. The notations are chosen in such a way that Q is always the heat received by the system, so that $-Q$ is the heat given away by the system. In cyclic processes, the internal energy of the system does not change,

$$\Delta U = \oint dU = 0. \tag{1.10.2}$$

Thus, integrating Eq. (1.6.1) over the cycle, one obtains

$$0 = \oint (\delta Q + \delta W) = Q + W = Q_2 + Q_1 + W, \tag{1.10.3}$$

so that the work done by the system is $-W = Q_2 + Q_1$. Inserting this into Eq. (1.10.1), one obtains

$$\eta = 1 + \frac{Q_1}{Q_2}. \tag{1.10.4}$$

One can see that $\eta < 1$. To make the efficiency η as high as possible, one should minimize Q_1.

Let us consider a particular process, the Carnot cycle, the working body being an ideal gas. The Carnot cycle consists of two isotherms, T_1 and T_2, and two adiabats connecting them (see Fig. 1.3). The cycle goes in the clockwise direction, so that the work done by the

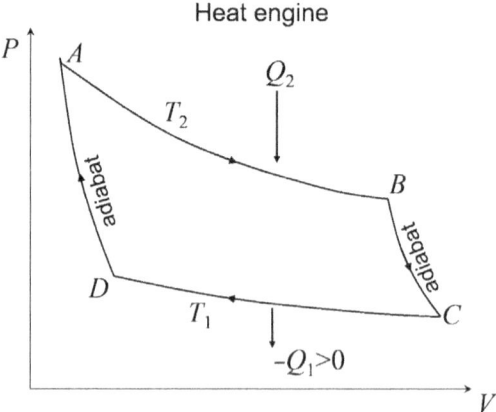

Figure 1.3.   Carnot cycle of a heat engine.

system is positive, $-W = \oint PdV > 0$. One can see that the heat Q_2 is received on the isothermal path AB at $T = T_2$, whereas the heat $-Q_1 > 0$ is given away on the isothermal path CD at $T = T_1$. There is no heat exchange on the adiabatic paths BC and DA. This is why the Carnot cycle is the most simple and thus the most fundamental of thermodynamic cycles. To calculate Q_1 and Q_2, one can use the fact that for the ideal gas, $U = U(T)$, so that along the isotherms $U = $ const., and thus, $dU = \delta Q + \delta W = 0$. The work in the isothermal process has been calculated above. Using Eq. (1.5.4), one obtains

$$Q_2 = Q_{AB} = -W_{AB} = -\nu RT_2 \ln \frac{V_A}{V_B} = \nu RT_2 \ln \frac{V_B}{V_A} > 0,$$

$$Q_1 = Q_{CD} = -W_{CD} = -\nu RT_1 \ln \frac{V_C}{V_D} < 0. \qquad (1.10.5)$$

Now, with the help of the adiabat equations, one can show that the logarithms in these expressions are the same. Indeed, from Eq. (1.8.8) follows

$$T_2 V_B^{\gamma-1} = T_1 V_C^{\gamma-1},$$

$$T_2 V_A^{\gamma-1} = T_1 V_D^{\gamma-1}. \qquad (1.10.6)$$

Dividing these equations by each other, one obtains $V_B/V_A = V_C/V_D$. Thus, the ratio Q_1/Q_2 simplifies as

$$\frac{Q_1}{Q_2} = -\frac{T_1}{T_2}, \qquad (1.10.7)$$

and Eq. (1.10.4) gives the famous Carnot formula

$$\eta = 1 - \frac{T_1}{T_2}. \qquad (1.10.8)$$

One can see that the efficiency η becomes close to 1 if the temperature of the cold reservoir is close to absolute zero. Practically, it is impossible to realize. In standard engines, T_1 is the temperature at normal conditions, $T_1 = 300$ K. Then the temperature T_2 of the hot reservoir must essentially exceed T_1. In particular, for $T_2 = 600$ K, one obtains $\eta = 0.5$. In practice, the processes in heat engines deviate from the Carnot cycle, which can be shown to be optimal. This leads to a further decrease in the efficiency η.

Refrigerator or heat pump

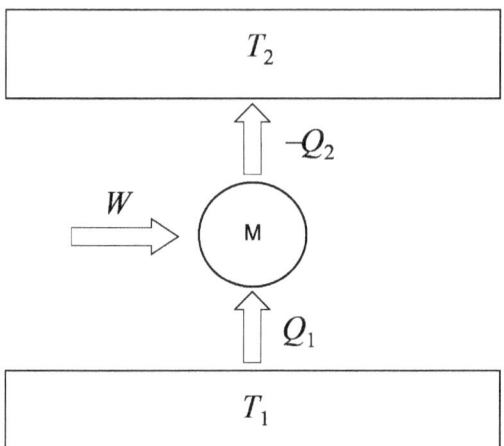

Figure 1.4. Schematic representation of a refrigerator (environment at T_2) or heat pump (environment at T_1).

1.10.2   Refrigerator

Refrigerators are inverted heat engines, see Fig. 1.4. The work is done on the system that extracts the heat Q_1 from the cold reservoir (eventually lowering its temperature) and gives heat $-Q_2 > 0$ to the hot reservoir (the environment). The efficiency of the refrigerator is defined as the energy output Q_1 to the energy input, W. As the energy balance of the refrigerator is similar to that of the engine, $W + Q_1 + Q_2 = 0$, the efficiency formula becomes

$$c = \frac{Q_1}{W} = \frac{Q_1}{-Q_2 - Q_1}. \tag{1.10.9}$$

For the Carnot refrigerator, with the help of Eq. (1.10.7), this transforms to

$$c = \frac{T_1}{T_2 - T_1}. \tag{1.10.10}$$

One can see that the efficiency of a refrigerator can be very high if T_1 and T_2 are close. This is the situation when the refrigerator starts to work. However, as T_1 decreases well below T_2 (i.e., essentially the environmental temperature of $300\,\mathrm{K}$), the efficiency becomes small.

1.10.3 Heat pump

A heat pump is similar to a refrigerator, only the interpretation of reservoirs 1 and 2 changes. Now, 1 is the environment from where the heat is being pumped, whereas 2 is the reservoir that is being heated, e.g., a house. The efficiency of the heat pump is defined as the ratio of the output $-Q_2 > 0$ to the input $W = -Q_2 - Q_1$:

$$d = \frac{-Q_2}{W} = \frac{Q_2}{Q_2 + Q_1}. \qquad (1.10.11)$$

For the Carnot heat pump, the efficiency becomes

$$d = \frac{T_2}{T_2 - T_1}, \qquad (1.10.12)$$

the inverse of that of the Carnot engine. One can see that the efficiency of the heat pump is always greater than 1. If the temperatures T_1 and T_2 are close to each other, d becomes large. This characterizes the initial stage of the work of a heat pump. However, after T_2 increases well above T_1, the efficiency becomes close to 1. This is the regime of direct heating in which the energy input W is converted into heat, $-Q_2 > 0$, while the additional heat Q_1 obtained from the environment is small. Practically, we do not need to warm up our house to temperatures much greater than T_1. Typically, in winter, $T_1 \approx 270 \, \text{K}$, while $T_2 \approx 300 \, \text{K}$. In this case, $d \approx 10$. One can see that the heat pump is much more efficient than a direct heater. However, in reality, there are losses that lower its efficiency.

1.11 Entropy and thermodynamic potentials

From experiments, it follows that for any cyclic quasistatic process, the relation

$$\oint \frac{\delta Q}{T} = 0. \qquad (1.11.1)$$

is satisfied (although, in general, $\oint \delta Q \neq 0$). In fact, this result can be obtained from the second law of thermodynamics, see Section 1.14, but here, for the sake of simplicity, it is considered as a standalone experimental result at equilibrium. Equation (1.11.1) implies that its integrand is a differential of a function of state:

$$dS = \frac{\delta Q}{T}, \tag{1.11.2}$$

which has been called entropy S. Temperature derivatives of the entropy are related to heat capacities:

$$\left(\frac{\partial S}{\partial T}\right)_V = \frac{C_V}{T}, \quad \left(\frac{\partial S}{\partial T}\right)_P = \frac{C_P}{T}. \tag{1.11.3}$$

Inserting $\delta Q = TdS$ into the first law of thermodynamics, Eq. (1.6.1), one obtains the so-called *main thermodynamic identity*:

$$dU = TdS - PdV, \tag{1.11.4}$$

which is a differential of the internal energy as a function of two variables: $U = U(S,V)$. This means that the *natural* variables for the internal energy are entropy and volume.

Entropy provides a closure of the formalism of thermodynamics and allows us to obtain a number of important thermodynamic relations by calculus. In particular,

$$T = \left(\frac{\partial U}{\partial S}\right)_V, \quad -P = \left(\frac{\partial U}{\partial V}\right)_S. \tag{1.11.5}$$

The subscript S at the derivative means the adiabatic process. As the second mixed derivative does not depend on the order of differentiation, e.g.,

$$\frac{\partial^2 U}{\partial S \partial V} = \frac{\partial^2 U}{\partial V \partial S}, \tag{1.11.6}$$

one obtains the relation

$$\left(\frac{\partial T}{\partial V}\right)_S = -\left(\frac{\partial P}{\partial S}\right)_V. \tag{1.11.7}$$

Relations of this type in thermodynamics are called Maxwell relations.

Internal energy U expressed via S and V is one of the so-called *thermodynamic potentials*. The differential of a thermodynamic potential yields the first and second laws of thermodynamics combined. For $U(S,V)$, this is Eq. (1.11.4). One can introduce three other thermodynamic potentials, making the Legendre transformation with respect to the pairs of variables T, S or P, V.

The enthalpy H is defined by

$$H = U + PV, \qquad (1.11.8)$$

so that $dH = dU + PdV + VdP$. With the help of Eq. (1.11.4), one obtains

$$dH = TdS + VdP. \qquad (1.11.9)$$

One can see that the enthalpy should be considered as a function of S and P, which are its so-called *native variables*. One has

$$T = \left(\frac{\partial H}{\partial S}\right)_P, \quad V = \left(\frac{\partial H}{\partial P}\right)_S, \qquad (1.11.10)$$

and the Maxwell relation

$$\left(\frac{\partial T}{\partial P}\right)_S = \left(\frac{\partial V}{\partial S}\right)_P. \qquad (1.11.11)$$

The (Helmholtz) free energy F is defined by

$$F = U - TS, \qquad (1.11.12)$$

so that $dF = dU - TdS - SdT$. With the help of Eq. (1.11.4), one obtains

$$dF = -SdT - PdV. \qquad (1.11.13)$$

The native variables for F are T and V. One has

$$-S = \left(\frac{\partial F}{\partial T}\right)_V, \quad -P = \left(\frac{\partial F}{\partial V}\right)_T, \qquad (1.11.14)$$

and the Maxwell relation

$$\left(\frac{\partial S}{\partial V}\right)_T = \left(\frac{\partial P}{\partial T}\right)_V. \tag{1.11.15}$$

Finally, the Gibbs free energy is defined by

$$G = U - TS + PV. \tag{1.11.16}$$

The differential of G is given by

$$dG = -S\,dT + V\,dP, \tag{1.11.17}$$

so that $G = G(T, P)$ in native variables. One has

$$-S = \left(\frac{\partial G}{\partial T}\right)_P, \quad V = \left(\frac{\partial G}{\partial P}\right)_T, \tag{1.11.18}$$

and the Maxwell relation

$$-\left(\frac{\partial S}{\partial P}\right)_T = \left(\frac{\partial V}{\partial T}\right)_P. \tag{1.11.19}$$

Thermodynamic potentials are useful because they generate Maxwell relations that can be used to calculate thermodynamic coefficients that are difficult to measure, such as $(\partial S/\partial P)_T$ in Eq. (1.11.19). Also, one of the thermodynamic potentials can be calculated in statistical physics, and then all other thermodynamic properties follow from it. For instance, once F has been calculated, one can obtain the entropy from Eq. (1.11.14). After that, one obtains the internal energy as

$$U = F + TS = F - T\left(\frac{\partial F}{\partial T}\right)_V \tag{1.11.20}$$

and the heat capacity as

$$C_V = T\left(\frac{\partial S}{\partial T}\right)_V = -T\left(\frac{\partial^2 F}{\partial T^2}\right)_V. \tag{1.11.21}$$

On the other hand, the second of Eq. (1.11.14) yields the equation of state.

1.11.1 Volume dependence of the internal energy

Let us now derive the formula for the volume derivative of the internal energy, Eq. (1.7.14). Dividing the main thermodynamic identity, Eq. (1.11.4), by dV while keeping $T = \text{const.}$, one obtains

$$\left(\frac{\partial U}{\partial V}\right)_T = T\left(\frac{\partial S}{\partial V}\right)_T - P. \qquad (1.11.22)$$

Then, using the Maxwell relation, Eq. (1.11.15), one arrives at

$$\left(\frac{\partial U}{\partial V}\right)_T = T\left(\frac{\partial P}{\partial T}\right)_V - P, \qquad (1.11.23)$$

Equation (1.7.14) quoted above. From the equation of state of the ideal gas follows $\left(\frac{\partial U}{\partial V}\right)_T = 0$.

1.11.2 Entropy of the ideal gas

The entropy of the ideal gas $S(T, V)$ can be found by integrating the relations

$$\left(\frac{\partial S}{\partial T}\right)_V = \frac{C_V}{T}, \quad \left(\frac{\partial S}{\partial V}\right)_T = \left(\frac{\partial P}{\partial T}\right)_V = \frac{\nu R}{V}. \qquad (1.11.24)$$

For the perfect gas, $C_V = \text{const.}$, integration of the first relation over T yields $S(T, V) = C_V \ln T + f(V)$, where $f(V)$ is the integration constant in the temperature integration that can depend on V. To define the latter, one can use the second relation above, which becomes $df/dV = \nu R/V$. Integration of it yields $f = \nu R \ln V + S_0$. Finally, one obtains

$$S = C_V \ln T + \nu R \ln V + S_0 = C_V \ln \left(TV^{\gamma-1}\right) + S_0. \qquad (1.11.25)$$

This formula defines the entropy up to an arbitrary constant S_0. In the adiabatic process of a perfect gas, $TV^{\gamma-1} = \text{const.}$ and the entropy does not change.

Consider an experiment: A container is separated into two parts by a membrane. In one part of the container is gas and in the other part is vacuum. The membrane is suddenly removed, after which the gas spreads over the whole container. This is an irreversible process that leads from one equilibrium state to the other. The gas does not do work, and it also does not receive heat. Thus, in this process, its internal energy is conserved. Since $U = U(T)$ for the ideal gas, its temperature does not change as well. The entropy in the new equilibrium state is larger due to the increase in volume. For the perfect gas, this is described by Eq. (1.11.25). At the moment when the membrane was removed, the old equilibrium state had become a nonequilibrium state. Spreading the gas over the whole container is the process of relaxation, i.e., approaching the equilibrium. In such irreversible processes, the entropy increases, and this increase is not related to the heat, unlike the case of reversible processes, Eq. (1.11.2).

For the free energy, combining Eqs. (1.8.15), (1.11.12), and (1.11.25) yields

$$F = C_V T + U_0 - C_V T \ln \left(TV^{\gamma-1}\right) - T S_0. \qquad (1.11.26)$$

One can check that Eqs. (1.11.20) and (1.11.21) yield the familiar results for the perfect gas. On the other hand, from Eq. (1.11.14), with the use of Eq. (1.8.7), follows the equation of state

$$P = -\left(\frac{\partial F}{\partial V}\right)_T = \frac{C_V T (\gamma - 1)}{V} = \frac{\nu RT}{V}. \qquad (1.11.27)$$

1.11.3 Adiabatic thermodynamic coefficients

In Section 1.4, we considered the thermodynamic coefficient relating changes of two thermodynamic quantities with the third quantity fixed. In particular, we discussed the isothermal compressibility $\varkappa_T$. If the system is thermally isolated or compression is so fast, so that the heat exchange that requires some time is effectively switched off, the compression process is adiabatic. One can define the adiabatic compressibility similar to Eq. (1.4.2):

$$\varkappa_S \equiv -\frac{1}{V}\left(\frac{\partial V}{\partial P}\right)_S. \qquad (1.11.28)$$

To express $\varkappa_S$ through experimentally measurable quantities, one can use Eq. (1.4.1) in which $dT \neq 0$ since T changes in the adiabatic process. First, one has to work out the condition $S = \text{const}$. One can try writing

$$dS = \left(\frac{\partial S}{\partial V}\right)_P dV + \left(\frac{\partial S}{\partial P}\right)_V dP \qquad (1.11.29)$$

and setting $dS = 0$, to obtain the relation between dV and dP, needed in Eq. (1.11.28). Unfortunately, the derivatives of the entropy above cannot be readily expressed via experimentally measurable quantities. Using the pair of variables (T, V), one can write

$$dS = \left(\frac{\partial S}{\partial T}\right)_V dT + \left(\frac{\partial S}{\partial V}\right)_T dV. \qquad (1.11.30)$$

Setting $dS = 0$ yields the relation between dT and dV, which with the help of Eqs. (1.11.3) and (1.11.15) becomes

$$0 = \frac{C_V}{T}dT + \left(\frac{\partial P}{\partial T}\right)_V dV. \qquad (1.11.31)$$

Similar, using the pair of variables (T, P), one can write

$$dS = \left(\frac{\partial S}{\partial T}\right)_P dT + \left(\frac{\partial S}{\partial P}\right)_T dP. \qquad (1.11.32)$$

Upon setting $dS = 0$ and using Eqs. (1.11.3) and (1.11.19), one obtains

$$0 = \frac{C_P}{T}dT - \left(\frac{\partial V}{\partial T}\right)_P dP. \qquad (1.11.33)$$

Now, eliminating dT from Eqs. (1.11.31) and (1.11.33), one obtains

$$-\frac{1}{C_V}\left(\frac{\partial P}{\partial T}\right)_V dV = \frac{1}{C_P}\left(\frac{\partial V}{\partial T}\right)_P dP. \qquad (1.11.34)$$

This yields

$$\varkappa_S \equiv -\frac{1}{V}\left(\frac{\partial V}{\partial P}\right)_S = \frac{1}{\gamma V}\left(\frac{\partial V}{\partial T}\right)_P\left(\frac{\partial T}{\partial P}\right)_V, \qquad (1.11.35)$$

where $\gamma \equiv C_P/C_V$. Finally, using the triple product rule, Eq. (1.4.6), one arrives at the relation

$$\varkappa_S = \frac{\varkappa_T}{\gamma}, \quad \varkappa_T \equiv -\frac{1}{V}\left(\frac{\partial V}{\partial P}\right)_T. \tag{1.11.36}$$

Since $C_P > C_V$ for all substances, $\varkappa_S < \varkappa_T$ is universally valid.

This formula can be checked on the ideal gas for which the adiabat equation in the P, V variables has the form $PV^\gamma = $ const., Eq. (1.8.9). Writing it in the form $\ln P + \gamma \ln V = $ const. and differentiating, one obtains

$$\frac{dP}{P} + \gamma\frac{dV}{V} = 0, \tag{1.11.37}$$

where

$$\left(\frac{\partial V}{\partial P}\right)_S = -\frac{1}{\gamma}\frac{V}{P} \tag{1.11.38}$$

and

$$\varkappa_S = \frac{1}{\gamma P} = \frac{\varkappa_T}{\gamma}, \tag{1.11.39}$$

see Eq. (1.4.8). The isothermal susceptibility can also be derived by this method, using only the isotherm $PV = $ const. rather than the full equation of state.

Adiabatic compression usually causes the temperature to increase. The corresponding thermodynamic coefficient can be obtained from Eq. (1.11.31):

$$\left(\frac{\partial T}{dV}\right)_S = -\frac{T}{C_V}\left(\frac{\partial P}{\partial T}\right)_V = -\frac{T}{C_V}\frac{\beta}{\varkappa_T}, \tag{1.11.40}$$

where Eq. (1.4.7) was used. For the ideal gas, this formula yields

$$\left(\frac{\partial T}{dV}\right)_S = -\frac{T}{C_V}\frac{\nu R}{V} = -\frac{P}{C_V}. \tag{1.11.41}$$

Another adiabatic coefficient can be obtained from Eq. (1.11.33):

$$\left(\frac{\partial T}{dP}\right)_S = \frac{T}{C_P}\left(\frac{\partial V}{\partial T}\right)_P = \frac{VT\beta}{C_P}. \tag{1.11.42}$$

For the ideal gas, this formula yields

$$\left(\frac{\partial T}{dP}\right)_S = \frac{T}{C_P}\frac{\nu R}{P} = \frac{V}{C_P}. \tag{1.11.43}$$

1.12 Problems

1.12.1 Entropy change in the isobaric–isochoric cycle of an ideal gas

Show that the entropy change in the cyclic process of an ideal gas, which is represented by a rectangle in the (P,V) diagram, is zero (Fig. 1.5).

Solution: In the isobaric process of an ideal gas, the infinitesimal amount of heat is given by

$$\delta Q = dU + PdV = C_V dT + PdV.$$

From the equation of state of the ideal gas

$$PV = \nu RT$$

follows

$$T = \frac{PV}{\nu R}, \quad dT = \frac{PdV}{\nu R}$$

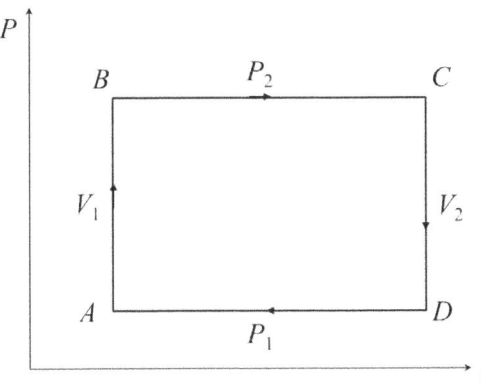

Figure 1.5. Isobar–isochore cycle.

Substituting this into

$$dS = \frac{\delta Q}{T},$$

one obtains

$$dS = \frac{C_V P dV/(\nu R) + P dV}{PV/(\nu R)} = (C_V + \nu R)\frac{dV}{V} = C_P \frac{dV}{V}.$$

In the isochoric process of the ideal gas, δQ is given by

$$\delta Q = C_V dT = C_V \frac{V dP}{\nu R},$$

thus

$$dS = \frac{\delta Q}{T} = C_V \frac{V dP}{PV} = C_V \frac{dP}{P}.$$

In our cyclic process,

$$\Delta S_{AB} = \int_{P_1}^{P_2} C_V \frac{dP}{P} = C_V \ln \frac{P_2}{P_1} > 0,$$

$$\Delta S_{CD} = C_V \ln \frac{P_1}{P_2} = -\Delta S_{AB},$$

$$\Delta S_{BC} = \int_{V_1}^{V_2} C_P \frac{dV}{V} = C_P \ln \frac{V_2}{V_1} > 0,$$

$$\Delta S_{DA} = C_P \ln \frac{V_1}{V_2} = -\Delta S_{BC}.$$

The total entropy change

$$\Delta S = \Delta S_{AB} + \Delta S_{BC} + \Delta S_{CD} + \Delta S_{DA} = 0,$$

as it should be.

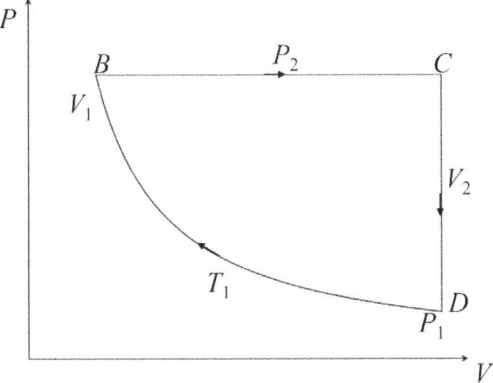

Figure 1.6. Isobar–isochore–isotherm cycle.

1.12.2 Entropy change in the isobaric–isochoric–isothermic cycle of an ideal gas

Show that the entropy change in the cyclic process of an ideal gas that includes an isobar, an isochore, and an isotherm is zero (Fig. 1.6).

Solution: Using the results of the solution of the previous problem, one finds

$$\Delta S_{BC} = C_P \ln \frac{V_2}{V_1} > 0, \quad \Delta S_{CD} = C_V \ln \frac{P_1}{P_2} < 0.$$

In the isothermal process of an ideal gas, $dU = 0$, thus $\delta Q = PdV$ and

$$dS = \frac{\delta Q}{T} = \frac{PdV}{T} = \nu R \frac{dV}{V}.$$

This yields

$$\Delta S_{DB} = \nu R \ln \frac{V_1}{V_2} < 0.$$

Using the equation of state of the ideal gas, on the ends of the isotherm, one has

$$\frac{P_1}{P_2} = \frac{V_1}{V_2} \quad \Rightarrow \quad \Delta S_{CD} = C_V \ln \frac{V_1}{V_2}.$$

The total entropy change over the cycle is

$$\Delta S = \Delta S_{BC} + \Delta S_{CD} + \Delta S_{DB} = (C_P - C_V - \nu R) \ln \frac{V_2}{V_1} = 0,$$

as it should be.

1.12.3 Entropy of a perfect gas

Calculate the entropy of a perfect gas as a function of (V, T) by integration using $S = \int \delta Q / T$.

Solution: Define $S(V_0, T_0) = S_0$ as a reference point and calculate the entropy $S(V, T)$ via the integral of $\delta Q / T$ over a path $(V_0, T_0) \Rightarrow (V, T)$, i.e.,

$$S(V, T) = S_0 + \int_{(V_0, T_0)}^{(V, T)} \frac{\delta Q}{T}.$$

As the entropy is a state function, its value does not depend on the path. Thus, one can choose the most convenient path, for instance, $(V_0, T_0) \Rightarrow (V_0, T) \Rightarrow (V, T)$. At the first stage, only the temperature is changing while the work is zero, thus

$$\delta Q = dU = \left(\frac{\partial U}{\partial T} \right)_V dT = C_V dT.$$

Integration for the perfect gas ($C_V = $ const.) proceeds as follows:

$$S(V_0, T) = S_0 + \int_{T_0}^T \frac{C_V dT}{T} = S_0 + C_V \ln \frac{T}{T_0}.$$

At the second stage, $T = $ const. As for the ideal gas $U = U(T)$, it does not change, and δQ is only due to work:

$$\delta Q = P dV.$$

Using the equation of state of the ideal gas $PV = \nu RT$, this can be rewritten as

$$\delta Q = \nu RT \frac{dV}{V}.$$

Now, integration with $T = \text{const.}$ proceeds as follows:

$$S(V,T) = S(V_0, T) + \nu R \int_{V_0}^{V} \frac{dV}{V} = S_0 + C_V \ln \frac{T}{T_0} + \nu R \ln \frac{V}{V_0}.$$

Here, the terms with T_0 and V_0 can be absorbed in the constant:

$$S(V,T) = C_V \ln T + \nu R \ln V + \text{const.}$$

Using $C_P - C_V = \nu R$ (Meyer's relation) and $\gamma = C_P/C_V$, one can rewrite this formula as

$$S(V,T) = C_V (\ln T + (\gamma - 1) \ln V) + \text{const.} = C_V \ln T V^{\gamma - 1} + \text{const.}$$

The argument of the logarithm is constant in the adiabatic process, $S = \text{const.}$, thus the result has an expected behavior and passes an error check.

1.12.4 Internal energy of a perfect gas in natural variables

Express the energy of a perfect gas in the natural variables, $U = U(S, V)$, and check relations

$$T = \left(\frac{\partial U}{\partial S} \right)_V, \quad -P = \left(\frac{\partial U}{\partial V} \right)_S, \quad \left(\frac{\partial T}{\partial V} \right)_S = - \left(\frac{\partial P}{\partial S} \right)_V.$$

Solution: In the V, T variables, the energy of a perfect gas has the form

$$U = C_V T,$$

where a constant has been dropped for simplicity. The entropy of the perfect gas is given by

$$S = C_V \ln T V^{\gamma - 1}, \tag{1.12.1}$$

where again a constant has been dropped. From here, one can express T as a function of S:

$$T = \frac{1}{V^{\gamma - 1}} \exp \left(\frac{S}{C_V} \right). \tag{1.12.2}$$

Thus, the energy in its natural variables becomes

$$U(S, V) = \frac{C_V}{V^{\gamma-1}} \exp\left(\frac{S}{C_V}\right). \qquad (1.12.3)$$

Note that U depends on the volume!

Now, using

$$dU = TdS - PdV,$$

one can identify

$$T = \left(\frac{\partial U}{\partial S}\right)_V, \quad -P = \left(\frac{\partial U}{\partial V}\right)_S.$$

Let us check these relations. With the help of Eq. (1.12.2), one obtains

$$\left(\frac{\partial U}{\partial S}\right)_V = \frac{1}{V^{\gamma-1}} \exp\left(\frac{S}{C_V}\right) = T,$$

as it should be. Further, using $\gamma = C_P/C_V$ and $C_P - C_V = \nu R$, one obtains

$$\left(\frac{\partial U}{\partial V}\right)_S = -\frac{(\gamma - 1) C_V}{V^\gamma} \exp\left(\frac{S}{C_V}\right) = -\frac{\nu R}{V^\gamma} \exp\left(\frac{S}{C_V}\right).$$

Using Eq. (1.12.2),

$$\left(\frac{\partial U}{\partial V}\right)_S = -\frac{\nu RT}{V} = -\frac{PV}{V} = -P,$$

as it should be. Thus, we have obtained

$$P = \frac{\nu R}{V^\gamma} \exp\left(\frac{S}{C_V}\right) \qquad (1.12.4)$$

in the V, S variables. This formula could also be obtained from Eq. (1.12.1) and the equation of state similar to Eq. (1.12.2). Now, to check the Maxwell identity, using Eq. (1.12.2), one calculates

$$\left(\frac{\partial T}{\partial V}\right)_S = -\frac{\nu R}{C_V V^\gamma} \exp\left(\frac{S}{C_V}\right).$$

On the other hand, from Eq. (1.12.4), one obtains

$$-\left(\frac{\partial P}{\partial S}\right)_V = -\frac{\nu R}{C_V V^\gamma}\exp\left(\frac{S}{C_V}\right) = \left(\frac{\partial T}{\partial V}\right)_S,$$

as expected.

1.12.5 Thermodynamic potentials F and G of the perfect gas

Express thermodynamic potentials F and G of the perfect gas in terms of their natural variables and check relations similar to those in the preceding problem.

Solution: Using the definition of F and the formulas for U and S of a perfect gas, one obtains

$$F = U - TS = C_V T - TC_V \ln TV^{\gamma-1} = -C_V T \ln(TV^{\gamma-1}/e).$$
$$(1.12.5)$$

Since

$$dF = -SdT - PdV,$$

one can identify

$$-S = \left(\frac{\partial F}{\partial T}\right)_V, \quad -P = \left(\frac{\partial F}{\partial V}\right)_T.$$

The entropy follows from Eq. (1.12.5) as

$$S = -\left(\frac{\partial F}{\partial T}\right)_V = -C_V + C_V + C_V \ln TV^{\gamma-1} = C_V \ln TV^{\gamma-1},$$

which is a known result. The pressure is

$$P = -\left(\frac{\partial F}{\partial V}\right)_T = \left(\frac{\partial(C_V T \ln V^{\gamma-1})}{\partial V}\right)_T = C_V T(\gamma - 1)\left(\frac{\partial(\ln V)}{\partial V}\right)_T$$
$$= \frac{C_V T(\gamma - 1)}{V} = \frac{\nu RT}{V},$$

also a known result.

The Maxwell relation

$$\left(\frac{\partial S}{\partial V}\right)_T = \left(\frac{\partial P}{\partial T}\right)_V$$

is now checked as follows:

$$\left(\frac{\partial S}{\partial V}\right)_T = \frac{\partial}{\partial V} C_V \ln TV^{\gamma-1} = C_V (\gamma - 1) \frac{\partial \ln V}{\partial V} = \frac{\nu R}{V}.$$

On the other hand,

$$\left(\frac{\partial P}{\partial T}\right)_V = \frac{\nu R}{V} = \left(\frac{\partial S}{\partial V}\right)_T,$$

as expected.

For the Gibbs thermodynamic potential G, all calculations are parallel to those for F, only one has to express all the formulas via P instead of V, using the equation of state of the ideal gas.

1.12.6 Thermodynamics from F

The Helmholtz free energy of a certain gas has the form

$$F = -\frac{\nu^2 a}{V} - \nu RT \ln (V - \nu b) + J(T).$$

Find the equation of state of this gas as well as its internal energy, entropy, heat capacities C_P and C_V and, in particular, their difference $C_P - C_V$.

Solution: To find the equation of state, one has to find P that will be a function of the native variables V, T:

$$P = -\left(\frac{\partial F}{\partial V}\right)_T = -\frac{\nu^2 a}{V^2} + \frac{\nu RT}{V - \nu b}. \tag{1.12.6}$$

Rearranging this formula, one obtains

$$\left(P + \frac{\nu^2 a}{V^2}\right)(V - \nu b) = \nu RT, \tag{1.12.7}$$

the van der Waals equation of a nonideal gas.

Next, the entropy is given by

$$S = -\left(\frac{\partial F}{\partial T}\right)_V = \nu R \ln(V - \nu b) - J'(T).$$

Now, the internal energy becomes

$$U = F + TS = -\frac{\nu^2 a}{V} + J(T) - T J'(T).$$

The heat capacity C_V can be found as

$$C_V = \left(\frac{\partial U}{\partial T}\right)_V = -T J''(T)$$

or as

$$C_V = T\left(\frac{\partial S}{\partial T}\right)_V = -T J''(T).$$

Finding

$$C_P = T\left(\frac{\partial S}{\partial T}\right)_P$$

requires more work. An explicit way to do this is to express V in the form $V = V(P,T)$ everywhere with the help of Eq. (1.12.7). However, this V is a solution of a cubic equation that is better to avoid. Also, this method is inconvenient to study $C_P - C_V$ because both heat capacities have to be functions of the same variables. Thus, it is better to use the implicit method considering $S = S(V,T)$ but with $V = V(P,T)$. Then one obtains

$$C_P = T\left(\frac{\partial S}{\partial T}\right)_V + T\left(\frac{\partial S}{\partial V}\right)_T\left(\frac{\partial V}{\partial T}\right)_P = C_V + T\left(\frac{\partial S}{\partial V}\right)_T \Big/ \left(\frac{\partial T}{\partial V}\right)_P.$$

In this formula,

$$\left(\frac{\partial S}{\partial V}\right)_T = \frac{\nu R}{V - \nu b},$$

whereas

$$\left(\frac{\partial T}{\partial V}\right)_P = \frac{1}{\nu R}\frac{\partial}{\partial V}\left(P + \frac{\nu^2 a}{V^2}\right)(V - \nu b)$$

$$= \frac{1}{\nu R}\left[-\frac{2\nu^2 a}{V^3}(V - \nu b) + \left(P + \frac{\nu^2 a}{V^2}\right)\right].$$

Here, one can eliminate P using Eq. (1.12.6) that yields

$$\left(\frac{\partial T}{\partial V}\right)_P = \frac{1}{\nu R}\left[-\frac{2\nu^2 a}{V^3}(V - \nu b) + \frac{\nu RT}{V - \nu b}\right].$$

Gathering the terms, one obtains

$$C_P - C_V = \frac{(\nu R)^2 T}{V - \nu b}\left/\left[-\frac{2\nu^2 a}{V^3}(V - \nu b) + \frac{\nu RT}{V - \nu b}\right]\right.,$$

further

$$C_P - C_V = \nu R\frac{\nu RT}{\nu RT - (2\nu^2 a/V^3)\left(V - \nu b\right)^2}$$

and, finally,

$$C_P - C_V = \frac{\nu R}{1 - \frac{2\nu^2 a(V - \nu b)^2}{\nu RTV^3}} > \nu R.$$

One can see that at high temperatures and large volumes, the additional term in the denominator becomes small and Meyer's relation for the ideal gas arises.

1.13 The third law of thermodynamics

Analyzing the experimental data, Nerst has concluded that in the limit $T \to 0$, the entropy becomes a constant independent of other thermodynamic parameters, such as volume and pressure:

$$\left(\frac{\partial S}{\partial V}\right)_{T\to 0} = \left(\frac{\partial S}{\partial P}\right)_{T\to 0} = 0. \tag{1.13.1}$$

Since in thermodynamics entropy is defined up to a constant, Planck has suggested defining

$$S(T \to 0) = 0. \tag{1.13.2}$$

Explanation of these results is possible only within statistical physics. It turns out that statistically defined entropy always satisfies Eq. (1.13.1), whereas Eq. (1.13.2) holds for most substances. Some materials have a degenerate ground state, and in this case, the entropy tends to a finite constant at $T \to 0$.

Let us consider the consequences of the third law. First, integrating the first equation of Eq. (1.11.3), one obtains

$$S = \int_0^T \frac{C_V}{T} dT + S_0. \tag{1.13.3}$$

If C_V is finite at $T \to 0$, the entropy logarithmically diverges, which contradicts the third law. Thus,

$$C_V(T \to 0) = 0. \tag{1.13.4}$$

The same condition for C_P can be proven in a similar way. Note that the divergence of the entropy of the ideal gas, Eq. (1.11.25), at $T \to 0$ only proves that the concept of the ideal gas breaks down at low temperatures, where gases become liquid and solid.

From Eq. (1.11.15) follows that the pressure thermal coefficient vanishes at absolute zero:

$$\left(\frac{\partial P}{\partial T}\right)_V\bigg|_{T\to 0} = 0. \tag{1.13.5}$$

From Eq. (1.11.19) follows that the thermal expansion coefficient vanishes:

$$\left(\frac{\partial V}{\partial T}\right)_P\bigg|_{T\to 0} = 0. \tag{1.13.6}$$

1.13.1   Open systems and chemical potential

In the above considerations, the mass of the system was considered as constant and dropped from the arguments of thermodynamic functions. If the system can exchange mass with the environment or if there are chemical reactions in the system, the mass or masses of components change and can cause changes in other quantities. For instance, if mass is added to the system with a constant volume, the pressure typically increases. In the sequel, we will consider the number of particles N instead of the mass or the number of kilomoles ν. The connection between N and ν has been discussed above, see Eq. (1.3.3) and the following. Using N is preferable in statistical physics, while ν is more convenient in chemistry.

With account of mass changes, the internal energy becomes a function of three variables, $U = U(S, V, N)$, and the main thermodynamic identity, Eq. (1.11.4), should be modified as follows:

$$dU = TdS - PdV + \mu dN. \tag{1.13.7}$$

Here,

$$\mu = \left(\frac{\partial U}{\partial N}\right)_{S,V} \tag{1.13.8}$$

is the chemical potential per particle. One can also use chemical potential per kilomole by writing $\mu d\nu$ in Eq. (1.13.7). For a multi-component system such as a mixture of different gases described by numbers of particles N_i, one has to replace $\mu dN \Rightarrow \sum_i \mu_i dN_i$.

Equation (1.13.8) looks nontrivial since special care should be taken to keep entropy constant while adding particles to the system. However, one can find a simpler representation for the chemical potential using the scaling argument. Since all arguments in $U(S, V, N)$ are extensive quantities, multiplying them all by a parameter λ means simply increasing the whole system by λ, which leads to the increase of U by λ. Mathematically, for any function f, this property can be expressed in the form

$$\lambda f(x, y, z) = f(\lambda x, \lambda y, \lambda z). \tag{1.13.9}$$

Differentiating this equation with respect to λ and then setting $\lambda \Rightarrow 1$ results in Euler's theorem,

$$f = x\frac{\partial f}{\partial x} + y\frac{\partial f}{\partial y} + z\frac{\partial f}{\partial z}. \tag{1.13.10}$$

For $U(S, V, N)$, the partial derivatives here are coefficients in Eq. (1.13.7), so that one obtains

$$U = TS - PV + \mu N. \qquad (1.13.11)$$

Using the definition of the Gibbs free energy, Eq. (1.11.16), one can rewrite this relation in the form

$$G = \mu N, \qquad (1.13.12)$$

i.e., the chemical potential is the Gibbs free energy per particle.

For open systems, the differentials of thermodynamic potentials, Eqs. (1.11.9), (1.11.13), and (1.11.17), are modified as follows:

$$dH = TdS + VdP + \mu dN, \qquad (1.13.13)$$
$$dF = -SdT - PdV + \mu dN, \qquad (1.13.14)$$
$$dG = -SdT + VdP + \mu dN. \qquad (1.13.15)$$

From Eq. (1.13.12) follows

$$dG = \mu dN + Nd\mu, \qquad (1.13.16)$$

which, combined with Eq. (1.13.15), yields

$$SdT - VdP + Nd\mu = 0, \qquad (1.13.17)$$

the Gibbs–Duhem equation.

1.13.2 The Ω-potential

Another thermodynamic potential for open systems is the Ω-potential that is sometimes convenient in statistical physics. Ω-potential is defined by

$$\Omega = -PV = F - G, \qquad (1.13.18)$$

expressed in terms of T, V, and μ as native variables. To obtain its differential, one can combine Eqs. (1.13.14) and (1.13.16) that yields

$$d\Omega = -SdT - PdV - Nd\mu. \qquad (1.13.19)$$

This formula implies

$$S = -\left(\frac{\partial \Omega}{\partial T}\right)_{V,\mu}, \quad P = -\left(\frac{\partial \Omega}{\partial V}\right)_{T,\mu} = -\frac{\Omega}{V}, \quad N = -\left(\frac{\partial \Omega}{\partial \mu}\right)_{T,V}.$$

$$(1.13.20)$$

If Ω is found, other thermodynamic potentials can be obtained from it. From Eq. (1.13.12) follows

$$G = -\mu \left(\frac{\partial \Omega}{\partial \mu}\right)_{T,V}. \tag{1.13.21}$$

Then, from Eq. (1.13.18), one obtains

$$F = \Omega - \mu \left(\frac{\partial \Omega}{\partial \mu}\right)_{T,V}. \tag{1.13.22}$$

For the internal energy from $U = F + TS$, one obtains

$$U = \Omega - \mu \left(\frac{\partial \Omega}{\partial \mu}\right)_{T,V} - T \left(\frac{\partial \Omega}{\partial T}\right)_{V,\mu}. \tag{1.13.23}$$

Finally, from $H = U + PV = U - \Omega$ follows

$$H = -\mu \left(\frac{\partial \Omega}{\partial \mu}\right)_{T,V} - T \left(\frac{\partial \Omega}{\partial T}\right)_{V,\mu}. \tag{1.13.24}$$

1.14 Second law of thermodynamics: Carnot's theorem

Experience shows that through a thermal contact between two bodies, the heat is always transmitted from the hotter body to the colder body, so that their temperatures equilibrate with time. This process cannot go in the opposite direction, although it would not contradict the first law of thermodynamics, the energy conservation. The statement of the impossibility of the heat transfer from the cold to the hot reservoir (that is not accompanied by work or its equivalent forms), constitutes the *second law* of thermodynamics.

After one has defined the entropy as a state function, one can look at the change of S in irreversible processes, e.g., processes of relaxation, just by comparing the entropies of the initial and final states.

As will be discussed in the following, as a result of irreversibility, the entropy of the system increases. This happens, in particular, in isolated systems, $\delta Q = 0$ and $\delta W = 0$, which approach equilibrium.

Using the second law of thermodynamics, one can prove the following famous Carnot's theorem stating that

> **the efficiency of any reversible heat engine operating between the heat reservoirs with temperatures T_1 and T_2 is equal to the efficiency of Carnot engine $\eta = 1 - T_1/T_2$,**

while the efficiency of any irreversible heat engine is lower than this. It should be noted that "operating between the heat reservoirs with T_1 and T_2" is a serious limitation on the heat engine that requires that the heat exchange takes place at constant temperatures T_1 and T_2 only, and thus, transition between states with these temperatures must be adiabatic. This excludes other types of cycles, e.g., the cycle in which adiabats are replaced by isochores on which the temperature changes. Effectively, the cycle performed by our "any" reversible heat engine is also a Carnot cycle. The only and a very serious difference with the above is that the working body is arbitrary and not necessarily an ideal gas.

To prove Carnot's theorem, one can let a general heat engine M′ drive Carnot machine M acting in reverse (i.e., as a refrigerator), see Fig. 1.7. Our composite machine does not do any work because $-W' = W$, thus the only effect is taking heat $Q_2' + Q_2$ from the hot reservoir and giving the same heat $-Q_1' - Q_1$ to the cold reservoir. Remember that the notations are chosen in such a way that Q is always the heat received by the system, so that $-Q$ is the heat given away by the system. Also, W is the work done on the system and $-W$ is the work done by the system. Consider the efficiencies of the two machines:

$$\eta' = \frac{-W'}{Q_2'}, \quad \eta = \frac{W}{-Q_2}$$

(we use the engine efficiency formula for both machines). Now, the difference of the efficiencies of the ideal Carnot engine M and our

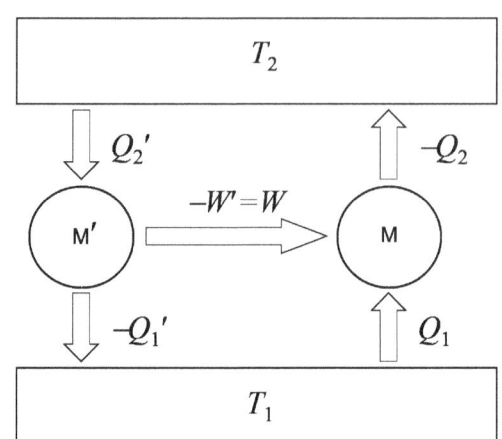

Figure 1.7. Coupled engine M' and Carnot engine M. The former is driving the latter, acting as a refrigerator.

engine M' becomes

$$\eta' - \eta = \frac{-W'}{Q_2'} + \frac{W}{Q_2} = \frac{W}{Q_2'} + \frac{W}{Q_2} = \frac{W(Q_2 + Q_2')}{Q_2 Q_2'}. \qquad (1.14.1)$$

According to the *second law* of thermodynamics, $Q_2' + Q_2 \geq 0$, and by definition, $Q_2 < 0$. Thus, $\eta' - \eta \leq 0$, which proves Carnot's theorem. In the reversible case, no heat flows from the hot to the cold reservoir, and thus, $\eta' = \eta$.

Now, we rewrite the condition above using Eq. (1.10.4) for M' (arbitrary body) and Eq. (1.10.8) for M (ideal gas!). This yields

$$\eta' - \eta = 1 + \frac{Q_1'}{Q_2'} - \left(1 - \frac{T_1}{T_2}\right) = \frac{Q_1'}{Q_2'} + \frac{T_1}{T_2} \leq 0. \qquad (1.14.2)$$

Since $Q_2' > 0$, this can be rewritten as

$$\frac{Q_1'}{T_1} + \frac{Q_2'}{T_2} \leq 0. \qquad (1.14.3)$$

Next, for an infinitesimally narrow Carnot cycle, made on the arbitrary body, discarding primes, one obtains

$$\frac{\delta Q_1}{T_1} + \frac{\delta Q_2}{T_2} \leq 0. \qquad (1.14.4)$$

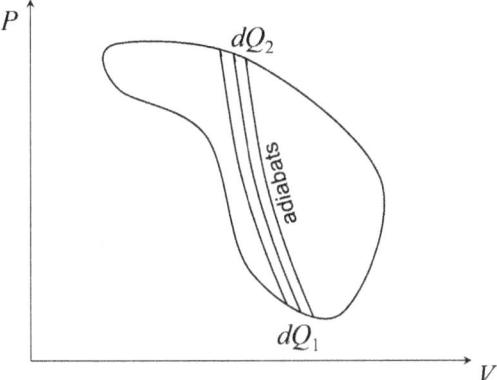

Figure 1.8. An arbitrary cycle represented as a series of infinitesimal Carnot cycles.

An infinite number of such Carnot cycles can be used to represent an arbitrary cyclic process with T changing arbitrarily, as the long adiabats cancel each other, see Fig. 1.8. As a result, one obtains

$$\oint \frac{\delta Q}{T} \leq 0, \tag{1.14.5}$$

the famous Clausius inequality. For any reversible cycle, there is the equality sign in this formula, as one has in Eq. (1.11.1). This means that there is a state function, entropy, defined by Eq. (1.11.2).

Let us consider the change of entropy in irreversible processes. In nonequilibrium states, the thermodynamic entropy is undefined. If, however, the initial and final states of an irreversible process are equilibrium states, the entropy in these states is defined, so one can define the entropy change $\Delta S_{12} = S_2 - S_1$. One can always find a reversible process connecting 1 and 2, the so-called *equivalent reversible process*. Both processes can be joined into an irreversible cyclic process, for which Eq. (1.14.5) applies and takes the form

$$\int_1^2 \frac{\delta Q}{T} + \int_2^1 \frac{\delta Q_{\text{reversible}}}{T} \leq 0. \tag{1.14.6}$$

Since the reversible integral is related to the change of the entropy, one obtains

$$\int_1^2 \frac{\delta Q_{\text{reversible}}}{T} = \Delta S_{12} \geq \int_1^2 \frac{\delta Q}{T}. \tag{1.14.7}$$

The differential form of this inequality reads

$$dS \geq \frac{\delta Q}{T}. \tag{1.14.8}$$

If the system is isolated, $\delta Q = \delta W = 0$. Still, the entropy of the system can increase, $dS \geq 0$, due to the irreversible processes inside the systems, e.g., relaxation to the equilibrium.

To illustrate the entropy increase in processes of approaching equilibrium, consider an isolated system that consists of two subsystems, each of them at internal equilibrium, whereas there is no equilibrium between the subsystems. Assume that the subsystems do not do work on each other and can only exchange heat. An example is two bodies with different temperatures brought into thermal contact with each other. Since both subsystems are at internal equilibrium, Eq. (1.11.2) is valid for both. Then it follows for the whole system that

$$dS = dS_1 + dS_2 = \frac{\delta Q_1}{T_1} + \frac{\delta Q_2}{T_2} = \delta Q_1 \left(\frac{1}{T_1} - \frac{1}{T_2} \right). \tag{1.14.9}$$

According to the second law of thermodynamics, the heat flows from the hot to the cold body, $T_1 < T_2$ and $\delta Q_1 > 0$. This yields $dS \geq 0$, although the whole system is isolated, $\delta Q = \delta Q_1 + \delta Q_2 = 0$.

1.14.1 Main thermodynamic relation with irreversibility

An important comment to Eq. (1.11.4) is in order. Since the employed expressions for the heat and work are reversible, one could think that Eq. (1.11.4) is valid for reversible processes only. In the literature, one can find the inequality $dU \leq TdS - PdV$, where $<$ is said to be realized for irreversible processes, where $\delta Q < TdS$. However, for basic systems that are completely described by the minimal set of thermodynamic variables such as P and V, Eq. (1.11.4) is exact, and it does not rely on the process being reversible. Since all quantities in this equation are state functions, it simply gives a small difference of the internal energy between two close points on the plot S, V. The inequality sign in $dU \leq TdS - PdV$ can only be achieved if there are irreversible processes in the system, such as dissolution, chemical reactions, and phase transformations, which are described by additional parameters.

For the basic thermodynamic system, both Eqs. (1.6.1) and (1.11.4) are exact. There is no contradiction with $\delta Q < TdS$ for irreversible processes because in these cases, one has $\delta W > -PdV$ as well. An example is free expansion of a gas into vacuum (Section 1.11.2), where $\delta Q = \delta W = 0$, but $TdS = PdV > 0$. Generally, for irreversible processes, one can write

$$\delta Q = \delta Q_{\text{reversible}} + \delta Q_{\text{dissipative}}, \quad \delta Q_{\text{reversible}} = TdS,$$

$$\delta W = \delta W_{\text{reversible}} + \delta W_{\text{dissipative}}, \quad \delta W_{\text{reversible}} = -PdV.$$
$$(1.14.10)$$

Then from

$$\delta Q + \delta W = \delta Q_{\text{reversible}} + \delta W_{\text{reversible}} = TdS - PdV \qquad (1.14.11)$$

follows $\delta Q_{\text{dissipative}} + \delta W_{\text{dissipative}} = 0$, and further,

$$\delta Q + \delta W_{\text{dissipative}} = TdS. \qquad (1.14.12)$$

This shows that the dissipative work is equivalent to the heat supply. $\delta W_{\text{dissipative}} \geq 0$ restoring the balance in Eq. (1.14.8) can always be found in particular irreversible processes described by a single set of thermodynamic quantities. One example is stirring a liquid. Another example is free expansion of the ideal gas, see the comment below Eq. (1.11.25).

1.15 Entropy maximum, thermodynamic equilibrium, phase transitions

1.15.1 Heat exchange

When two bodies with temperatures T_1 and T_2 are brought into thermal contact, the heat flows from the hot body to the cold body, so that the temperatures equilibrate. This is the second law of thermodynamics that follows from experiments. It was shown above (see Eq. (1.14.9)), that the total entropy $S = S_1 + S_2$ in the process of equilibration (i.e., relaxation) increases. When the equilibrium is reached, S should attain its maximal value. Investigation of the behavior of the total entropy near its maximum is the subject of this chapter.

Consider first the simplest case in which the two bodies do not exchange mass and keep their volumes unchanged. Then the transferred heat is the only source of the energy change, $dU_1 = \delta Q_1$ and $dU_2 = \delta Q_2$. Since the system of two bodies is isolated from the rest of the world, $\delta Q_1 + \delta Q_2 = 0$, and thus, $dU_1 + dU_2 = 0$. The corresponding changes of the entropies up to the second order in the transferred energy are given by

$$dS_1 = \left(\frac{\partial S_1}{\partial U_1}\right)_V dU_1 + \frac{1}{2}\left(\frac{\partial^2 S_1}{\partial U_1^2}\right)_V (dU_1)^2 \qquad (1.15.1)$$

and similarly for dS_2. For the derivatives, one has

$$\left(\frac{\partial S}{\partial U}\right)_V = \frac{1}{T},$$

$$\left(\frac{\partial^2 S}{\partial U^2}\right)_V = -\frac{1}{T^2}\left(\frac{\partial T}{\partial U}\right)_V = -\frac{1}{T^2 C_V}. \qquad (1.15.2)$$

Eliminating $dU_2 = -dU_1$, one obtains

$$dS = dS_1 + dS_2 = \left(\frac{1}{T_1} - \frac{1}{T_2}\right) dU_1$$

$$-\frac{1}{2}\left(\frac{1}{T_1^2 C_{V1}} + \frac{1}{T_2^2 C_{V2}}\right)(dU_1)^2. \qquad (1.15.3)$$

One can see that the extremum of S corresponds to $T_1 = T_2$, the thermal equilibrium. The quadratic term in this formula shows that this extremum is a maximum, provided the heat capacities are positive:

$$C_V > 0. \qquad (1.15.4)$$

The latter is the condition of thermodynamic stability. The state with $C_V < 0$ would be unstable as the heat flow from the hot to the cold body would lead to an increase of the differences in their temperatures instead of equilibration. The initial state with $T_1 = T_2$ would be unstable with respect to the transfer of a small amount of energy between them as a result of a fluctuation that would lead to an avalanche-like further transfer of energy in the same direction. Indeed, for $C_V < 0$, the temperature on the receiving side would decrease, and according to the second law of thermodynamics, the heat will flow to this side, further decreasing its temperature. As $C_P > C_V$, one concludes that C_P is also positive. Equation (1.15.4) complements the condition of mechanical stability, Eq. (1.4.3).

At equilibrium, $T_1 = T_2 = T$, Eq. (1.15.3) becomes

$$dS = -\frac{1}{2T^2}\left(\frac{1}{C_{V1}} + \frac{1}{C_{V2}}\right)(dU_1)^2. \tag{1.15.5}$$

If the second body is much larger than the first one, it can be considered as a bath. In this case, $C_{V2} \gg C_{V1}$ and the second fraction in the above formula can be neglected. Using $dU_1 = C_{V1}dT_1$ and dropping the index 1 for the bathed system, one obtains

$$dS = -\frac{C_V}{2T^2}(dT)^2. \tag{1.15.6}$$

This formula gives the entropy decrease caused by the deviation of the system's temperature by a small amount dT from the bath temperature T.

1.15.2 General case of thermodynamic equilibrium

Let us now consider two systems in contact that can exchange energy, volume, and mass. Exchanging volume means that there is a movable membrane between the two bodies, so that the bodies can do work on each other. Exchanging mass means that this membrane is penetrable by particles. Resolving Eq. (1.13.7) for dS, one obtains, to the first order,

$$dS_1 = \frac{1}{T_1}dU_1 + \frac{P_1}{T_1}dV_1 - \frac{\mu_1}{T_1}dN_1 \tag{1.15.7}$$

and a similar expression for dS_2. Here, one could include second-order terms like those in Eq. (1.15.1) to find extended conditions of stability. Using the constraints

$$dU_1 + dU_2 = 0, \qquad dV_1 + dV_2 = 0, \qquad dN_1 + dN_2 = 0, \tag{1.15.8}$$

one obtains for the total entropy change

$$dS = \left(\frac{1}{T_1} - \frac{1}{T_2}\right)dU_1 + \left(\frac{P_1}{T_1} - \frac{P_2}{T_2}\right)dV_1 - \left(\frac{\mu_1}{T_1} - \frac{\mu_2}{T_2}\right)dN_1. \tag{1.15.9}$$

Carnot's theorem proves that it follows from the experimental fact (or postulate) of the heat flow from the hot to the cold body that the entropy of an isolated system can only increase. This statement turns out to be more general than the initial observation. Indeed, the entropy also increases in the case of free expansion of a gas, see Eq. (1.11.25). Here, the number of gas molecules in different regions of space equilibrate rather than the temperature. The law of the entropy increase mandates that the free expansion process goes in the direction of expansion and not in the direction of compression. It is noteworthy that we do not have to adopt the irreversibility of the free expansion as an additional postulate because it already follows from the existing postulate of the temperature equilibration. Thus, we require that $dS \geq 0$ in Eq. (1.15.9). This has three consequences:

(i) The energy flows from the hotter body to the colder body.
(ii) The body with a higher pressure expands at the expense of the body with lower pressure.
(iii) The particles diffuse from the body with a higher chemical potential μ to that with the lower μ.

The thermodynamic equilibrium is characterized by

$$T_1 = T_2 \quad \text{(thermal equilibrium)}, \qquad (1.15.10)$$
$$P_1 = P_2 \quad \text{(mechanical equilibrium)}, \qquad (1.15.11)$$
$$\mu_1 = \mu_2 \quad \text{(diffusive equilibrium)}. \qquad (1.15.12)$$

Next, the total entropy should have a maximum with respect to all three variables at the equilibrium. Investigating this requires adding second-order terms to Eq. (1.15.9), as was done in the case of the pure heat exchange in the previous section. The analysis is somewhat cumbersome, but the results can be figured out. First, the condition of thermal stability, Eq. (1.15.4), should be satisfied. Second, the condition of mechanical stability, Eq. (1.4.3), should be satisfied. Third, the diffusive stability condition should exist to the effect that adding particles to the system at constant volume and internal energy should increase its chemical potential.

This condition has yet to be worked out, but one can expect that it follows from the other two conditions, i.e., it is not the third independent condition.

1.15.3 Phase transitions

The results for the diffusive equilibrium in the preceding section can be applied to phase transitions. If different phases of the same substance, such as solid, liquid, and gas, are in contact, particles can migrate from one phase to the other across the phase boundary. In this case, the phase with a higher chemical potential recedes and the phase with a lower chemical potential grows. The phase boundary moves across the sample until the receding phase disappears completely. In basic thermodynamic systems, the chemical potentials of the phases μ_i depend on P and T, so that for a given set P, T, the phase with the lowest chemical potential will be realized. Transitions between the phases occur at $\mu_i(P, T) = \mu_j(P, T)$, which describes lines in the P, T diagram. Phase transitions of this kind are called *first-order* phase transitions. In the first-order phase transitions, phases are labeled by discrete variables i, such as 1 for solid, 2 for liquid, and 3 for gas.

Another kind of phase transition is the *second-order* phase transition. In the second-order phase transitions, phases are described by the so-called order parameter (say, η) that is zero in one of the phases and nonzero in the other phase. Most of the second-order transitions are controlled by the temperature, and $\eta = 0$ in the high-temperature (symmetric) phase. As the temperature is lowered across the phase transition point T_c, the order parameter continuously grows from zero, usually as $\eta \propto (T_c - T)^\beta$ with a *critical index* $\beta > 0$ close to T_c. The temperature dependence of the order parameter and other thermodynamic quantities is *singular* at T_c, unlike the situation in first-order transitions. An example of a second-order transition is ferromagnetic ordering below the Curie point T_c in iron and other ferromagnetic materials.

Phase transformation in second-order phase transitions can be formally considered in the same way as above. If one obtains the chemical potential in the form $\mu(\eta)$, one can consider boundaries between regions with different values of η and thus of μ. Then the particles will migrate from the phase with the higher μ to that with the lower μ, so that the spatial boundary between the phases will move until the equilibrium state with the lowest μ is established. Moreover, since η can change continuously, η can adjust in a uniform way without any phase boundaries in the whole system, decreasing its chemical potential everywhere. In isolated systems, the number of particles is conserved, so that the minimum of μ implies a minimum of the Gibbs potential G, see Eq. (1.13.12). If pressure and volume are irrelevant in the problem, such as in the case of magnetism, G is the same as the free energy F, so this is F that has to be minimized with respect to the order parameter η. Practically, if one finds F (or G) as a function of the order parameter η, one can determine the equilibrium value of η from

$$\partial F(\eta)/\partial \eta = 0. \tag{1.15.13}$$

An example of this scheme is a mean-field theory of ferromagnetism considered in Chapter 3, on statistical mechanics.

Chapter 2

Molecular Theory of Ideal Gases

Molecular theory can be considered as a preliminary to statistical physics. While the latter employs a more sophisticated formalism encompassing quantum-mechanical systems, the former operates with a classical ideal gas. Molecular theory studies the relation between the temperature of the gas and the kinetic energy of the molecules, pressure on the walls due to the impact of the molecules, distribution of molecules over velocities, etc. All these results can be obtained in a more formal way in statistical physics. However, it is convenient to first study molecular theory at a more elementary level. On the other hand, molecular theory includes the *kinetics* of classical gases that studies nonequilibrium phenomena such as *heat conduction* and *diffusion* (not a part of this course).

2.1   Basic assumptions of the molecular theory

1. Motion of atoms and molecules is described by classical mechanics.
2. The number of particles in a considered macroscopic volume is very large. As there are about 10^{19} molecules in 1 cm^3 at normal conditions, this assumption holds down to high vacuums. Due to the large number of particles, the impacts of individual particles on the walls merge into time-independent pressure.
3. The characteristic distance between the molecules largely exceeds the molecular size and the typical radius of intermolecular forces. This assumption allows us to consider the gas as ideal,

with the internal energy dominated by the kinetic energy of the molecules. In describing equilibrium properties of the ideal gas, collisions between the molecules can be neglected. If pressure is increased and/or temperature is decreased, this assumption becomes violated, and the gas becomes nonideal and then condenses into a liquid or solid.

4. The molecules are uniformly distributed within the container. In most cases, this is true. In the case of strong enough potential fields, this condition may be violated. For instance, the density (concentration) of molecules in the atmosphere decreases with the height. This decrease is slow, however, so that in a laboratory-size container, the gas is still approximately uniform.

5. Directions of velocities of the molecules are uniformly distributed. This is the hypothesis of the *molecular chaos* that is always true.

An important independent consideration is that the motion of molecules along different perpendicular directions such as x, y, and z is independent. Indeed, if a molecule experiences a force impulse (e.g., because of a collision) in the direction x, it does not change its velocity components v_y and v_z. This leads to the postulate of factorization of their distribution function into parts depending on the velocity components v_x, v_y, and v_z. The latter allows finding this distribution function that turns out to be exponentially dependent on the molecules' kinetic energy.

2.2 Characteristic lengths of the gas

The concentration of molecules n is defined by

$$n \equiv \frac{N}{V}, \tag{2.2.1}$$

where V is the volume of the container and N is the total number of molecules. If the concentration is nonuniform, one has to modify this formula by considering a small volume around a particular point in the space that contains a macroscopic number of molecules.

The characteristic distance r_0 between the molecules can be estimated as

$$r_0 = \frac{1}{n^{1/3}}. \tag{2.2.2}$$

Note that one cannot introduce a meaningful average distance between the molecules as the latter will be dominated by the molecules far apart.

Let a be the radius of the molecule or an atom. (For simplicity, we consider them as spheres.) Then the third assumption requires $a \ll r_0$. There are also long-range attractive forces between the molecules, but they are weak and do not essentially deviate molecular trajectories from straight lines if the temperature is high enough.

One can define the mean free path l of the molecules as the distance they typically travel before colliding with another molecule. As the molecule is moving straight, it will hit any molecule whose center is located at a distance less than $2a$ from the molecule's trajectory, i.e., any molecule within a cylinder of radius $2a$. The cross-section of this cylinder is $\sigma = \pi(2a)^2$. The free path ends, on average, when the height l of the cylinder reaches such a value that there is one molecule within the cylinder: $l\sigma n = 1$. Thus, one obtains the expression for l and important inequalities:

$$l = \frac{1}{\sigma n} \sim \frac{1}{a^2 n} = \left(\frac{r_0}{a}\right)^2 r_0 \gg r_0 \gg a. \tag{2.2.3}$$

Although the numerical factor in l can be calculated more accurately, one can already see that l is very large in the ideal gas, so that one can neglect collisions on the way to the wall in calculating the pressure and other quantities.

2.3 Velocity distribution function of molecules

Whereas the distribution of molecules in space is practically uniform, their distribution in the space of velocities (v_x, v_y, v_z) is nontrivial. One can introduce the distribution function $G(v_x, v_y, v_z)$ via

$$dN = NG(v_x, v_y, v_z)dv_x dv_y dv_z, \tag{2.3.1}$$

where dN is the number of molecules with the velocities within the elementary volume

$$dv_x dv_y dv_z \equiv d^3v \equiv d\mathbf{v} \tag{2.3.2}$$

around the velocity vector specified by its components (v_x, v_y, v_z). Integration over the whole velocity space gives the total number of molecules N, thus $G(v_x, v_y, v_z)$ satisfies the normalization condition

$$1 = \int \int \int_{-\infty}^{\infty} dv_x dv_y dv_z G(v_x, v_y, v_z). \tag{2.3.3}$$

As the directions of the molecular velocities are distributed uniformly, $G(v_x, v_y, v_z)$ in fact depends only on the absolute value of the velocity, the speed

$$v = \sqrt{v_x^2 + v_y^2 + v_z^2}. \tag{2.3.4}$$

Using the expression for the elementary volume in the spherical coordinates

$$d^3v = dv \times v d\theta \times v \sin\theta d\varphi = v^2 dv \sin\theta d\theta d\varphi, \tag{2.3.5}$$

one can rewrite Eq. (2.3.1) as

$$dN = NG(v)v^2 dv d\Omega, \tag{2.3.6}$$

where

$$d\Omega \equiv \sin\theta d\theta d\varphi \tag{2.3.7}$$

is the elementary body angle. The number of molecules within the spherical shell of width dv can be obtained by integration over irrelevant directions. Since for the area of a sphere of unit radius, one has

$$\int_{\text{sphere}} d\Omega = \int_0^{\pi} d\theta \sin\theta \int_0^{2\pi} d\varphi = 2\pi \int_{-1}^{1} dx = 4\pi \tag{2.3.8}$$

$(x \equiv \cos\theta)$, the number of molecules within the spherical shell becomes

$$dN = NG(v)4\pi v^2 dv = Nf(v)dv. \qquad (2.3.9)$$

Here, we have introduced the distribution function over molecular speeds

$$f(v) = 4\pi v^2 G(v), \qquad (2.3.10)$$

which is also normalized by 1:

$$1 = \int_0^\infty dv f(v). \qquad (2.3.11)$$

Now, Eq. (2.3.6) can be rewritten in terms of $f(v)$ as

$$dN = Nf(v)dv\frac{d\Omega}{4\pi}. \qquad (2.3.12)$$

It is remarkable that the functional form of $G(v)$ or $f(v)$ can be found from symmetry arguments only. This will be done later. Here, we introduce the *average* speed

$$\bar{v} = \int_0^\infty dv\, vf(v) \qquad (2.3.13)$$

and the *mean square* speed

$$\overline{v^2} = \int_0^\infty dv\, v^2 f(v). \qquad (2.3.14)$$

The *root mean square* or *rms* speed is defined by

$$v_{\text{rms}} = \sqrt{\overline{v^2}}. \qquad (2.3.15)$$

The two characteristic speeds, $\bar{v}$ and v_{rms}, are of the same order of magnitude and differ only by a numerical factor of order 1. We will see that the molecular flux is proportional to $\bar{v}$, while the pressure on the walls is proportional to $\overline{v^2}$.

2.4   Molecular flux

Molecular flux Φ is defined as the number of molecules dN crossing a unit surface in one direction during a unit of time. For instance, molecular flux determines the rate of molecules striking the wall or exiting the container through a small orifice in the wall (effusion). The expression for the flux reads

$$\Phi = \frac{dN}{dSdt}, \qquad (2.4.1)$$

where dS is the elementary surface. As the molecules are approaching the surface dS from all directions θ, φ of a hemisphere, one should first consider the number of molecules $dN_{\theta,\varphi}$ coming from a particular direction θ, φ within the body angle $d\Omega$ around it. From these molecules, we single out molecules with the speeds in the interval dv around v, thus obtaining $dN_{v,\theta,\varphi}$. The latter is the number of molecules within the slant cylinder with the base area dS and height $v\cos\theta dt$, see Fig. 2.1. The volume of this cylinder is $dV = dSv\cos\theta dt$, and the total number of molecules in it is $dN_V = ndV$, where the concentration n is defined by Eq. (2.2.1). From this total number of molecules dN_V, one has to pick those

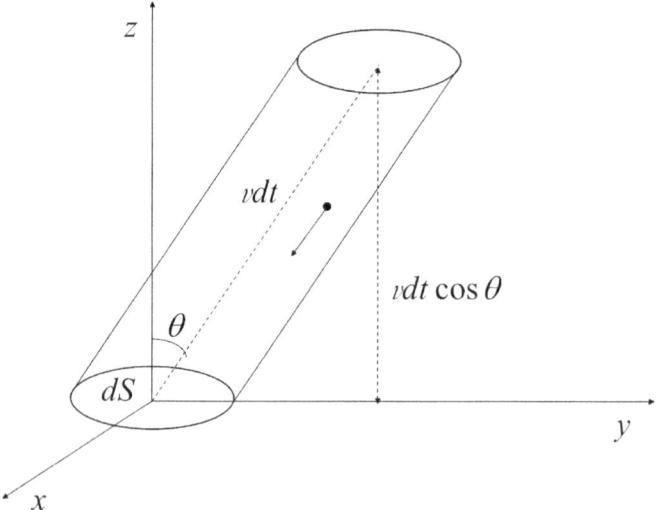

Figure 2.1.   Slant cylinder used in the calculation of the molecular flux. Here, $\varphi = \pi/2$.

within the given velocity interval specified by v, θ, φ. With the help of Eq. (2.3.12), with $N \Rightarrow dN_V$, this yields

$$dN_{v,\theta,\varphi} = dN_V f(v)dv\frac{d\Omega}{4\pi} = ndSv\cos\theta dt f(v)dv\frac{\sin\theta d\theta d\varphi}{4\pi}. \quad (2.4.2)$$

Integrating over v, θ, φ and using Eqs. (2.3.13) and (2.4.1), one obtains the flux

$$\Phi = \int \frac{dN_{v,\theta,\varphi}}{dSdt} = n\int_0^\infty dv\, vf(v)\frac{1}{4\pi}\int_0^{\pi/2} d\theta\sin\theta\cos\theta\int_0^{2\pi} d\varphi$$

$$= n\times\bar{v}\times\frac{1}{4\pi}\int_0^1 dx\, x\times 2\pi. \quad (2.4.3)$$

Note that the molecules are approaching the wall only from inside of the container, $0 \leq \theta \leq \pi/2$. After calculating the integral over x, one obtains

$$\Phi = \frac{1}{4}n\bar{v}. \quad (2.4.4)$$

If, instead of the wall, one considers the flux through a flat region on the area dS inside the container, the molecules will approach this area from both directions, so that one has to integrate over the interval $0 \leq \theta \leq \pi$ that results in a zero flux (the number of molecules crossing in both directions is the same).

2.5 Gas pressure on the walls

As said above, the gas pressure is due to the impact of molecules on the walls. Considering the elementary surface dS as above, one can define pressure as $P = dF/dS$, where dF is the force acting upon the surface dS from the molecules. The force can be obtained from Newton's second law in the form $dp/dt = F$, where p is the momentum of the molecules that changes in time due to the rebound from the wall. Adopting this to our case yields the formula

$$P = \frac{dp}{dSdt}, \quad (2.5.1)$$

where dp is the change of the momentum of the molecules within the slant cylinder considered in the preceding section.

The change of the momentum of a single molecule in the collision with the wall is, strictly speaking, not well defined. The problem is that at the atomic level, walls are rough and the incident molecule can rebound in different directions. On the other hand, one can consider the realistic rough wall as built of small plates oriented at different angles. Since the pressure will not depend on the orientation of the elementary surfaces (similar to Pascal's law), one can dismiss the effect of the wall roughness. Another effect that can make the calculation more involved is the inelasticity of the molecule–wall collision due to the exchange of energy between the molecules and the atoms of the wall. As a result, some molecules rebound stronger and some rebound weaker relative to the elastic collision. A detailed analysis shows that the molecule–wall collisions average out to the elastic collision if the walls and the gas have the same temperatures and thus are at equilibrium. Thus, here we will consider the collisions of molecules with the wall as elastic collisions with a flat surface.

As the change of the momentum of an individual molecule in an elastic collision is given by $mv \cos \theta - (-mv \cos \theta) = 2mv \cos \theta$, m being the mass of a molecule, similar to Eq. (2.4.3), one obtains

$$
\begin{aligned}
P &= \int 2mv \cos \theta \frac{dN_{v,\theta,\varphi}}{dS dt} \\
&= 2nm \int_0^\infty dv \, v^2 f(v) \frac{1}{4\pi} \int_0^{\pi/2} d\theta \sin \theta \cos^2 \theta \int_0^{2\pi} d\varphi \\
&= 2nm \times \overline{v^2} \times \frac{1}{4\pi} \int_0^1 dx \, x^2 \times 2\pi
\end{aligned} \tag{2.5.2}
$$

that results in

$$
P = \frac{1}{3} nm \overline{v^2}. \tag{2.5.3}
$$

2.6 Molecular interpretation of temperature and equipartition of energy

Rewriting Eq. (2.5.3) as $PV = (1/3)Nm\overline{v^2}$ and comparing this with the equation of state in the form $PV = Nk_B T$, one obtains

$$
k_B T = \frac{1}{3} m \overline{v^2}, \tag{2.6.1}
$$

the fundamental relation between the temperature and average kinetic energy of the molecule $\bar{\varepsilon}$. This relation can be rewritten in the form

$$\bar{\varepsilon} = \frac{1}{2}m\overline{v^2} = \frac{3}{2}k_BT. \tag{2.6.2}$$

Since $\overline{v^2} = \overline{v_x^2} + \overline{v_y^2} + \overline{v_z^2}$ and by symmetry $\overline{v_x^2} = \overline{v_y^2} = \overline{v_z^2} = \overline{v^2}/3$, for the kinetic energies corresponding to the three *degrees of freedom* x, y, z, one obtains

$$\bar{\varepsilon}_x = \bar{\varepsilon}_y = \bar{\varepsilon}_z = \frac{1}{2}k_BT, \tag{2.6.3}$$

i.e., the thermal energy per degree of freedom is $(1/2)k_BT$.

This is a particular case of the equipartition of energy valid for classical systems: There is thermal energy $(1/2)k_BT$ per degree of freedom. In addition to the three translational degrees of freedom, there are rotational and vibrational degrees of freedom if the molecules of the gas consist of more than one atom. Vibrational degrees of freedom are counted twice since there are both kinetic and potential energies involved.

For instance, for diatomic molecules, there are two rotational degrees of freedom corresponding to rotations around the two axes perpendicular to the axis connecting the two molecules. Also, there is one vibrational degree of freedom that is counted twice. The total number of degrees of freedom is

$$f = 3 + 2 + 2 = 7 \tag{2.6.4}$$

for diatomic molecules.

For multi-atomic molecules with $\mathcal{N} > 2$ atoms that are not aligned, there are three translational and three rotational degrees of freedom. The number of vibrational degrees of freedom is difficult to calculate directly. However, this number can be easily calculated by subtracting $3 + 3$ nonvibrational degrees of freedom from the total $3\mathcal{N}$ degrees of freedom. Thus, one obtains $3\mathcal{N} - 6$ vibrational degrees of freedom that should be counted twice. The total number of degrees of freedom for multi-atomic molecules is thus

$$f = 3 + 3 + 2(3\mathcal{N} - 6) = 6(\mathcal{N} - 1) \tag{2.6.5}$$

that yields $f = 12$ for $\mathcal{N} = 3$.

2.7 Heat capacity of the ideal gas

For the monoatomic gas, the average energy per particle is given by Eq. (2.6.2). Since there is no potential energy, the internal energy of the system is given by

$$U = \frac{3}{2} N k_B T. \tag{2.7.1}$$

Thus, the heat capacity at constant volume is

$$C_V = \left(\frac{\partial U}{\partial T} \right)_V = \frac{3}{2} N k_B. \tag{2.7.2}$$

Now, with the use of Mayer's formula, one obtains

$$C_P = C_V + N k_B = \frac{5}{2} N k_B \tag{2.7.3}$$

that yields $\gamma = C_P/C_V = 5/3$. For multi-atomic molecules, assuming equipartition results in $U = (f/2)N k_B T$ and

$$C_V = \frac{f}{2} N k_B, \qquad C_P = \frac{f+2}{2} N k_B, \qquad \gamma = 1 + \frac{2}{f}. \tag{2.7.4}$$

In all these cases, the heat capacity is a constant, so that the ideal gas is perfect gas. However, it turns out that vibrational degrees of freedom for multi-atomic gases are affected by quantum effects. As a result, these degrees of freedom are fully or partially frozen out, so that there is less thermal energy in them than the equipartition would suggest. Quantum effects are strongly pronounced at low temperatures, whereas at high temperatures, the vibrational modes behave classically. As a result, heat capacities increase with temperature, making the ideal gas not a perfect gas.

2.8 Maxwell–Boltzmann distribution function

In this section, the functional form of $G(v)$ will be found from symmetry arguments. First, the motion of molecules of an ideal gas along different axes x, y, z is completely independent, which implies statistical independence of different velocity components. Thus, $G(v)$ factorizes:

$$G(v) = G\left(\sqrt{v_x^2 + v_y^2 + v_z^2}\right) = g(v_x)g(v_y)g(v_z), \qquad (2.8.1)$$

the function g being normalized by 1:

$$1 = \int_{-\infty}^{\infty} dv_x g(v_x). \qquad (2.8.2)$$

This means that each velocity component has its own distribution function g. Indeed, the number of molecules within the shell dv_x around v_x is obtained by integrating Eq. (2.3.1) over irrelevant v_y, v_z:

$$dN = N\left[\int\!\!\int_{-\infty}^{\infty} dv_y dv_z G(v)\right] dv_x. \qquad (2.8.3)$$

With the help of Eqs. (2.8.1) and (2.8.2), this becomes

$$dN = Ng(v_x)dv_x, \qquad (2.8.4)$$

i.e., $g(v_x)$ is the distribution function of v_x.

Factorization of G and its spherical symmetry implemented in Eq. (2.8.1) already allow us to find its functional form. Taking the logarithm of this equation,

$$\ln G(v) = \ln g(v_x) + \ln g(v_y) + \ln g(v_z), \qquad (2.8.5)$$

and differentiating it with respect to v_x yields

$$\frac{G'(v)}{G}\frac{\partial v}{\partial v_x} = \frac{G'(v)}{G}\frac{v_x}{v} = \frac{g'(v_x)}{g(v_x)}. \qquad (2.8.6)$$

Rearranging and adding similar results for other components, one obtains

$$\frac{1}{v}\frac{G'(v)}{G} = \frac{1}{v_x}\frac{g'(v_x)}{g(v_x)} = \frac{1}{v_y}\frac{g'(v_y)}{g(v_y)} = \frac{1}{v_z}\frac{g'(v_z)}{g(v_z)}. \qquad (2.8.7)$$

Since different terms of these equations depend on different independent arguments, the only possibility to satisfy these equations is for all terms to be equal to the same constant:

$$\frac{1}{v}\frac{G'(v)}{G} = -2k, \qquad \frac{1}{v_x}\frac{g'(v_x)}{g(v_x)} = -2k, \qquad (2.8.8)$$

etc. Integrating these differential equations, one obtains

$$G(v) = Ae^{-kv^2}, \qquad (2.8.9)$$

where A is the integration constant. One can see that indeed $G(v)$ factorizes and

$$g(v_x) = A^{1/3}e^{-kv_x^2}. \qquad (2.8.10)$$

In fact, the derivation starting with Eq. (2.8.5) is probably unnecessary because Eq. (2.8.9) is the only factorizable function depending on v. One can persuade oneself of it through a long contemplation of Eq. (2.8.1).

Now, the two constants, k and A, can be found from the normalization condition, Eq. (2.3.11), and the condition for the mean square speed, Eq. (2.3.14), taking into account Eq. (2.6.1). As $f(v)$ is related to $G(v)$ by Eq. (2.3.10),

$$f(v) = 4\pi v^2 Ae^{-kv^2}, \qquad (2.8.11)$$

we will use the values of two Gaussian integrals

$$\int_0^\infty dx\, x^2 e^{-kx^2} = \frac{\sqrt{\pi}}{4}k^{-3/2},$$

$$\int_0^\infty dx\, x^4 e^{-kx^2} = \frac{3\sqrt{\pi}}{8}k^{-5/2}, \qquad (2.8.12)$$

which can be obtained by successive differentiation of the generic integral

$$\int_0^\infty dx\, e^{-kx^2} = \frac{\sqrt{\pi}}{2}k^{-1/2} \qquad (2.8.13)$$

with respect to k. The normalization condition, Eq. (2.3.11), works out as

$$1 = \int_0^\infty dv\, f(v) = 4\pi A \int_0^\infty dv\, v^2 e^{-kv^2} = \pi^{3/2} A k^{-3/2}. \qquad (2.8.14)$$

The condition for the mean square speed becomes

$$\frac{3k_BT}{m} = \overline{v^2} = \int_0^\infty dv\, v^2 f(v) = 4\pi A \int_0^\infty dv\, v^4 e^{-kv^2} = \frac{3\pi^{3/2}}{2} Ak^{-5/2}.$$

$$(2.8.15)$$

From these two equations, one finds

$$k = \frac{m}{2k_BT}, \qquad A = \left(\frac{k}{\pi}\right)^{3/2} = \left(\frac{m}{2\pi k_BT}\right)^{3/2}. \qquad (2.8.16)$$

Let us now write down the final results for the distribution functions. Equation (2.8.11) becomes

$$f(v) = \left(\frac{m}{2\pi k_BT}\right)^{3/2} 4\pi v^2 \exp\left(-\frac{\varepsilon}{k_BT}\right), \qquad \varepsilon = \frac{mv^2}{2}, \quad (2.8.17)$$

and Eq. (2.8.10) becomes

$$g(v_x) = \left(\frac{m}{2\pi k_BT}\right)^{1/2} \exp\left(-\frac{\varepsilon_x}{k_BT}\right), \qquad \varepsilon_x = \frac{mv_x^2}{2}. \qquad (2.8.18)$$

2.9 Characteristic speeds of gas molecules

Let us calculate the characteristic speeds for the ideal gas, two of which are the root-mean-square (rms) and average speed, Eqs. (2.3.13) and (2.3.15). The rms speed can be immediately obtained from Eq. (2.8.15):

$$v_{\rm rms} = \sqrt{\overline{v^2}} = \sqrt{\frac{3k_BT}{m}} \simeq 1.732\sqrt{\frac{k_BT}{m}}. \qquad (2.9.1)$$

To calculate the average speed, one makes use of the Gaussian integral

$$\int_0^\infty dx\, x^{2n+1} e^{-kx^2} = \frac{n!}{2k^{n+1}}, \qquad n = 0,1,2,\ldots, \qquad (2.9.2)$$

with $n = 1$. From Eq. (2.3.13), one obtains

$$\overline{v} = \int_0^\infty dv\, v f(v) = 4\pi A \int_0^\infty dv\, v^3 e^{-kv^2} = 4\pi A \frac{1}{2k^2} = \frac{2\pi}{k^2}\left(\frac{k}{\pi}\right)^{3/2}$$

$$= \frac{2}{\sqrt{\pi k}} = \sqrt{\frac{8k_BT}{\pi m}} \simeq 1.596\sqrt{\frac{k_BT}{m}}. \qquad (2.9.3)$$

The third characteristic speed is the most probable speed v_m corresponding to the maximum of $f(v)$. From

$$0 = \frac{d}{dv^2} v^2 e^{-kv^2} = e^{-kv^2} - v^2 k e^{-kv^2}, \qquad (2.9.4)$$

one obtains

$$v_m = \frac{1}{\sqrt{k}} = \sqrt{\frac{2k_B T}{m}} \simeq 1.414 \sqrt{\frac{k_B T}{m}}, \qquad (2.9.5)$$

the smallest of the three characteristic speeds.

2.10   Effusion

In Section 2.4, we have calculated the molecular flux Φ, the number of molecules hitting the unit area during the unit of time. If there is a small hole in the wall of the container, the molecules will escape through this hole. The process is called effusion. If the hole is small enough, it does not disturb the gas in the container close to the hole, and the result for the molecular flux given by Eq. (2.4.4) remains valid. Then the number of molecules leaving the container per second is given by $\Phi \Delta S$, where ΔS is the area of the hole.

One can ask what the speed distribution of escaping molecules. Certainly, this is not the Maxwell–Boltzmann distribution, already by the fact that the effusing molecules are moving all away from the container. Moreover, it turns out that the characteristic speeds of the effusing molecules are higher than the speeds of the molecules in the container. The reason is that faster molecules are approaching the hole from inside the container and exit at a higher rate than the slow molecules. There are quite a few very slow molecules in the flux through the hole.

One can obtain the speed distribution of the effusing molecules by removing the integration over v in Eq. (2.4.3). One can write

$$\Phi = \int_0^\infty dv\, \Phi_v, \qquad (2.10.1)$$

where

$$\Phi_v = \frac{1}{4} n v f(v) \qquad (2.10.2)$$

is the molecular flux corresponding to the speed interval dv around the value v. This is Φ_v that defines the speed distribution of the effusing molecules. Due to the additional v, this distribution is shifted to higher speeds. For instance, the most probable speed of the effusing molecules corresponds to the maximum of Φ_v and is defined as

$$0 = \frac{d}{dv} v^3 e^{-kv^2} = 3v^2 e^{-kv^2} - v^3 2kv e^{-kv^2}, \tag{2.10.3}$$

so that

$$v_{e,m} = \sqrt{\frac{3}{2k}} = \sqrt{\frac{3k_B T}{m}}. \tag{2.10.4}$$

One can see that $v_{e,m} > v_m$ given by Eq. (2.9.5). Similar, $v_{e,\mathrm{rms}} > v_{\mathrm{rms}}$ and $\bar{v}_e > \bar{v}$.

2.11 Problems

2.11.1 Factorizability of the distribution function

A hypothetical velocity distribution of an ideal gas has the form

$$G(v) = Ae^{-kv}.$$

Does $G(v)$ satisfy the molecular chaos postulate? Find A from the normalization condition. Find the most probable speed, average speed, and the rms speed. Find the distribution function $g(v_x)$. Check if G factorizes as $G(v) = g(v_x)g(v_y)g(v_z)$.

Solution: $G(v)$ satisfies the molecular chaos postulate because it does not depend on the directions of the velocities. The normalization condition for the distribution function is

$$1 = \int\int\int_{-\infty}^{\infty} dv_x dv_y dv_z G(v) = 4\pi \int_0^{\infty} v^2 dv G(v) = \int_0^{\infty} dv f(v).$$

Substituting the explicit form of $G(v)$, one obtains

$$1 = 4\pi A \int_0^{\infty} v^2 dv e^{-kv} = 4\pi A J_2(k),$$

where

$$J_2(k) = \int_0^{\infty} dv v^2 e^{-kv}.$$

Using

$$J_0(k) = \int_0^\infty dv e^{-kv} = \frac{1}{k},$$

one obtains

$$J_2(k) = \frac{d^2}{dk^2} J_0(k) = \frac{2}{k^3}.$$

Thus,

$$A = \frac{1}{4\pi J_2(k)} = \frac{k^3}{8\pi}.$$

The most probable speed v_m is defined by the maximum of $f(v)$, i.e.,

$$\max_v \left(v^2 e^{-kv} \right).$$

Taking the derivative over v, one obtains

$$0 = 2v e^{-kv} - kv^2 e^{-kv},$$

and thus,

$$v_m = \frac{2}{k}.$$

The average speed is given by

$$\bar{v} = \int_0^\infty dv\, v f(v) = 4\pi A \int_0^\infty dv\, v^3 e^{-kv} = \frac{k^3}{2} J_3(k).$$

Using

$$J_3(k) = -\frac{d}{dk} J_2(k) = \frac{6}{k^4},$$

one finally obtains

$$\bar{v} = \frac{3}{k}.$$

The average square speed is given by

$$\overline{v^2} = \int_0^\infty dv v^2 f(v) = 4\pi A \int_0^\infty dv v^4 e^{-kv} = \frac{k^3}{2} J_4(k).$$

Using

$$J_4(k) = -\frac{d}{dk} J_3(k) = \frac{24}{k^5},$$

one finally obtains

$$\overline{v^2} = \frac{12}{k^2}, \qquad (2.11.1)$$

and thus,

$$v_{\text{rms}} = \sqrt{\overline{v^2}} = \frac{2\sqrt{3}}{k}.$$

The distribution function for a single velocity component can be obtained by integrating $G(v)$ over the remaining velocity components

$$g(v_x) = \int\int_{-\infty}^\infty dv_y dv_z G\left(\sqrt{v_x^2 + v_y^2 + v_z^2}\right)$$

$$= 2\pi A \int_0^\infty v_\perp dv_\perp e^{-k\sqrt{v_x^2 + v_\perp^2}},$$

where $v_\perp = \sqrt{v_y^2 + v_z^2}$. Changing to the new variable $u = v_\perp^2$, one obtains

$$g(v_x) = \pi A \int_0^\infty du e^{-k\sqrt{v_x^2 + u}} = \frac{2\pi A}{k^2} e^{-k|v_x|}(1 + k|v_x|)$$

or, finally,

$$g(v_x) = \frac{k}{4} e^{-k|v_x|}(1 + k|v_x|).$$

Obviously,

$$G(v) \neq g(v_x)g(v_y)g(v_z),$$

i.e., our distribution function is not factorizable, and thus, it is not a good distribution function.

Still, let us investigate $g(v_x)$ obtained. Start with checking the normalization:

$$\int_{-\infty}^{\infty} dv_x g(v_x) = 2 \int_{0}^{\infty} dv_x g(v_x) = \frac{k}{2} \int_{0}^{\infty} dv_x e^{-k v_x} (1 + k v_x)$$

$$= \frac{k}{2} (J_0(k) + k J_1(k))$$

$$= \frac{k}{2} \left(\frac{1}{k} + k \frac{1}{k^2} \right) = 1,$$

as it should be. The average square velocity is

$$\overline{v_x^2} = \int_{-\infty}^{\infty} dv_x v_x^2 g(v_x) = \frac{k}{2} (J_2(k) + k J_3(k)) = \frac{k}{2} \left(\frac{2}{k^3} + k \frac{6}{k^4} \right) = \frac{4}{k^2}.$$

This is in accordance with Eq. (2.11.1) since $\overline{v_x^2} + \overline{v_y^2} + \overline{v_z^2} = \overline{v^2}$.

Chapter 3

Statistical Physics

Statistical physics considers systems of a large number of entities (particles), such as atoms, molecules, and spins. For these systems, it is impossible and does not even make sense to study the full microscopic dynamics. The only relevant information is, say, how many atoms have a particular energy, then one can calculate the observable thermodynamic values. That is, one has to know the distribution function of the particles over energies that defines the macroscopic properties. This gives the name *statistical physics* or statistical thermodynamics and defines the scope of this subject.

The approach outlined above can be used both at and off equilibrium. The branch of physics studying nonequilibrium situations is called *physical kinetics*. In this course, we study only statistical physics at equilibrium. It turns out that at equilibrium, the energy distribution function has an explicit general form, and the only problem is to calculate the observables. The term *statistical mechanics* means the same as statistical physics.

The formalism of statistical thermodynamics can be developed for both classical and quantum systems. The resulting energy distribution and calculation of observables are simpler in the classical case. However, the very formulation of the method is more transparent within the quantum mechanical formalism. In addition, the absolute value of the entropy, including its correct value at $T \to 0$, can only be obtained in the quantum case. To avoid double work, we will consider only quantum statistical thermodynamics in this course, limiting ourselves to systems without interaction. The more general quantum results will recover their classical forms in the classical limit.

3.1 Microstates and macrostates

It follows from quantum mechanics that the states of the system do not change continuously (like in classical physics) but are quantized. There is a huge number of discrete quantum states i with corresponding energy values ε_i being the main parameter characterizing these states. In the absence of interaction, each particle has its own set of quantum states which it can occupy. For identical particles, these sets of states are identical. One can think of boxes i into which particles are placed, N_i particles in the ith box. The particles can be distributed over the boxes in a number of different ways, corresponding to different *microstates*, in which the state i of each particle is specified. The information contained in the microstates is excessive, and the only meaningful information is provided by the numbers N_i that define the distribution of particles over their quantum states. These numbers N_i specify what in statistical physics is called *macrostate*. If these numbers are known, the energy and other quantities of the system can be found. It should be noted that the statistical macrostate also contains more information than the macroscopic physical quantities that follow from it, as a distribution contains more information than an average over it.

Each macrostate k, specified by the numbers N_i, can be realized by a number w_k of microstates, the so-called *thermodynamic probability*. The latter is typically a large number, unlike the usual probability that changes between 0 and 1. Redistributing the particles over the states i while keeping the same values of N_i generates different microstates within the same macrostate. The basic assumption of statistical mechanics is the equidistribution over microstates. That is, each microstate within a macrostate is equally probable for occupation. Macrostates having a larger w_k are more likely to be realized. As we will see, for large systems, thermodynamic probabilities of different macrostates vary in a wide range, and there is a state with the largest value of w that wins over all other macrostates.

If the number of quantum states i is finite, the total number of microstates can be written as

$$\Omega \equiv \sum_k w_k, \tag{3.1.1}$$

the sum rule for thermodynamic probabilities. For an isolated system, the number of particles N and the energy U are conserved, thus the numbers N_i satisfy the constraints

$$\sum_i N_i = N, \tag{3.1.2}$$

$$\sum_i N_i \varepsilon_i = U, \tag{3.1.3}$$

which limit the variety of the allowed macrostates k.

3.2 Two-state particles (coin tossing)

A tossed coin can land in two positions: head up or tail up. Considering the coin as a particle, one can say that this particle has two "quantum" states: 1 corresponding to the head and 2 corresponding to the tail. If N coins are tossed, this can be considered as a system of N particles with two quantum states each. The microstates of the system are specified by the states occupied by each coin. As each coin has two states, there are a total of

$$\Omega = 2^N \tag{3.2.1}$$

microstates. The macrostates of this system are defined by the numbers of particles in each state, N_1 and N_2. These two numbers satisfy the constraint condition (3.1.2), i.e., $N_1 + N_2 = N$. Thus, one can take, say, N_1 as the number k labeling macrostates. The number of microstates in one macrostate (i.e., the number of different microstates that belong to the same macrostate) is given by the binomial distribution

$$w_{N_1} = \frac{N!}{N_1!(N - N_1)!} = \binom{N}{N_1}. \tag{3.2.2}$$

This formula can be derived as follows. We have to pick N_1 particles to be in state 1. All others will be in state 2. How many ways are there to do this? The first "0" particle can be picked in N ways, the second one can be picked in $N - 1$ ways since we can choose from $N - 1$ particles only, the third "1" particle can be picked in $N - 2$ different ways, etc. Thus, one obtains the number of different ways to pick the particles is

$$N \times (N - 1) \times (N - 2) \times \cdots \times (N - N_1 + 1) = \frac{N!}{(N - N_1)!},$$

$$(3.2.3)$$

where the factorial is defined by

$$N! \equiv N \times (N - 1) \times \cdots \times 2 \times 1, \quad 0! = 1. \qquad (3.2.4)$$

The recurrent definition of the factorial is

$$N! = N (N - 1)!, \quad 0! = 1. \qquad (3.2.5)$$

The expression in Eq. (3.2.3) is not yet the thermodynamic probability w_{N_1} because it contains multiple counting of the same microstates. The realizations, in which N_1 "1" particles are picked in different orders, have been counted as different microstates, whereas they are the same microstate. To correct for the multiple counting, one has to divide the expression by the number of permutations $N_1!$ of the N_1 particles that yields Eq. (3.2.2). One can check that the condition (3.1.1) is satisfied:

$$\sum_{N_1=0}^{N} w_{N_1} = \sum_{N_1=0}^{N} \frac{N!}{N_1!(N - N_1)!} = 2^N. \qquad (3.2.6)$$

The thermodynamic probability w_{N_1} has a maximum at $N_1 = N/2$, half of the coins are heads and half of the coins are tails. This macrostate is the most probable state. Indeed, as for an individual coin, the probabilities to land head up and tail up are both equal to 0.5, this is what we expect. For large N, the maximum of w_{N_1} on N_1 becomes sharp.

To prove that $N_1 = N/2$ is the maximum of w_{N_1}, one can rewrite Eq. (3.2.2) in terms of the new variable $p = N_1 - N/2$ as

$$w_{N_1} = \frac{N!}{(N/2 + p)!(N/2 - p)!}. \tag{3.2.7}$$

One can see that w_p is symmetric around $N_1 = N/2$, i.e., $p = 0$. Using Eq. (3.2.5), one obtains

$$\frac{w_{N/2 \pm 1}}{w_{N/2}} = \frac{(N/2)!(N/2)!}{(N/2 + 1)!(N/2 - 1)!} = \frac{N/2}{N/2 + 1} < 1, \tag{3.2.8}$$

one can see that $N_1 = N/2$ is indeed the maximum of w_{N_1}. The binomial distribution is shown in Fig. 3.1 for three different values of N. As the argument, the variable $p/p_{\max} \equiv (N_1 - N/2)/(N/2)$ is used, so that one can put all the data on the same plot. One can see that in the limit of large N, the binomial distribution becomes narrow and centered at $p = 0$, i.e., $N_1 = N/2$. This practically means that if the coin is tossed many times, significant deviations from the 50:50 relation between the numbers of heads and tails will be extremely rare.

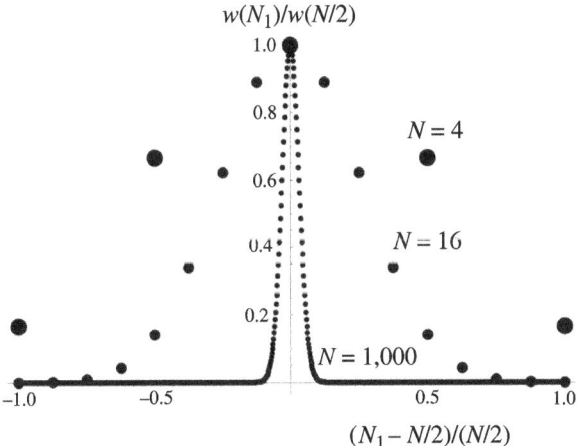

Figure 3.1. The binomial distribution for an ensemble of N two-state systems becomes narrow and peaked at $p \equiv N_1 - N/2 = 0$.

3.3 Stirling formula and thermodynamic probability at large N

Analysis of expressions with large factorials is simplified by the Stirling formula

$$N! \cong \sqrt{2\pi N} \left(\frac{N}{e}\right)^N. \tag{3.3.1}$$

In many important cases, the prefactor $\sqrt{2\pi N}$ is irrelevant, as we will see in the following. With the Stirling formula substituted, Eq. (3.2.7) becomes

$$
\begin{aligned}
w_{N_1} &\cong \frac{\sqrt{2\pi N}(N/e)^N}{\sqrt{2\pi \left(N/2 + p\right)} \left[(N/2 + p)/e\right]^{N/2+p}} \\
&\qquad \times \sqrt{2\pi \left(N/2 - p\right)} \left[(N/2 - p)/e\right]^{N/2-p} \\
&= \sqrt{\frac{2}{\pi N}} \frac{1}{\sqrt{1 - (\frac{2p}{N})^2}} \frac{N^N}{(N/2 + p)^{N/2+p} (N/2 - p)^{N/2-p}} \\
&= \frac{w_{N/2}}{\sqrt{1 - (\frac{2p}{N})^2}(1 + \frac{2p}{N})^{N/2+p}(1 - \frac{2p}{N})^{N/2-p}}, \tag{3.3.2}
\end{aligned}
$$

where

$$w_{N/2} \cong \sqrt{\frac{2}{\pi N}} 2^N \tag{3.3.3}$$

is the maximal value of the thermodynamic probability. Equation (3.3.2) can be expanded for $|p| \ll N$. Since p enters both the bases and the exponents, one has to be careful and expand the logarithm of w_{N_1} rather than w_{N_1} itself. The square root term in Eq. (3.3.2), can be discarded as it gives a negligible contribution of order p^2/N^2. One obtains

$$
\ln w_{N_1} \cong \ln w_{N/2} - \left(\frac{N}{2} + p\right) \ln \left(1 + \frac{2p}{N}\right) - \left(\frac{N}{2} - p\right) \ln \left(1 - \frac{2p}{N}\right)
$$

$$
\cong \ln w_{N/2} - \left(\frac{N}{2} + p\right) \left[\frac{2p}{N} - \frac{1}{2}\left(\frac{2p}{N}\right)^2\right]
$$

$$
- \left(\frac{N}{2} - p\right) \left[-\frac{2p}{N} - \frac{1}{2}\left(\frac{2p}{N}\right)^2\right]
$$

$$\cong \ln w_{N/2} - p - \frac{2p^2}{N} + \frac{p^2}{N} + p - \frac{2p^2}{N} + \frac{p^2}{N}$$

$$= \ln w_{N/2} - \frac{2p^2}{N}, \tag{3.3.4}$$

and thus,

$$w_{N_1} \cong w_{N/2} \exp\left(-\frac{2p^2}{N}\right). \tag{3.3.5}$$

One can see that w_{N_1} becomes very small if $|p| \gtrsim \sqrt{N}$, which for large N does not violate the applicability condition $|p| \ll N$. That is, w_{N_1} is small in the main part of the interval $0 \leq N_1 \leq N$ and is sharply peaked near $N_1 = N/2$.

3.4 Many-state particles

The results obtained in Section 3.2 for two-state particles can be generalized for n-state particles. We are looking for the number of ways to distribute N particles over n boxes, so that there are N_i particles in the ith box. That is, we look for the number of microstates in the macrostate described by the numbers N_i. The result is given by

$$w = \frac{N!}{N_1! N_2! \dots N_n!} = \frac{N!}{\prod_{i=1}^{n} N_i!}. \tag{3.4.1}$$

This formula can be obtained by using Eq. (3.2.2) successively. The number of ways to put N_1 particles in box 1 and the other $N - N_1$ in other boxes is given by Eq. (3.2.2). Then the number of ways to put N_2 particles in box 2 is given by a similar formula with $N \to N - N_1$ (there are only $N - N_1$ particles after N_1 particles have been put in box 1) and $N_1 \to N_2$. These number of ways should multiply. Then one considers box 3, etc., until the last box n. The resulting number of microstates is

$$w = \frac{N!}{N_1!(N - N_1)!} \times \frac{(N - N_1)!}{N_2!(N - N_1 - N_2)!} \times \frac{(N - N_1 - N_2)!}{N_3!(N - N_1 - N_2 - N_3)!} \times$$

$$\dots \times \frac{(N_{n-2} + N_{n-1} + N_n)!}{N_{n-2}!(N_{n-1} + N_n)!} \times \frac{(N_{n-1} + N_n)!}{N_{n-1}! N_n!} \times \frac{N_n!}{N_n! 0!}. \tag{3.4.2}$$

In this formula, all numerators except for the first one and all second terms in the denominators cancel each other, so that Eq. (3.4.1) follows.

3.5   Thermodynamic probability and entropy

Different macrostates k can be realized by varying numbers of microstates w_k. For large systems, $N \gg 1$, the difference between different thermodynamic probabilities w_k is tremendous, and there is a sharp maximum of w_k at some value of $k = k_{max}$. The main postulate of statistical physics is that in measurements on large systems, only the most probable macrostate, satisfying the constraints of Eqs. (3.1.2) and (3.1.3), makes a contribution. For instance, a macrostate of an ideal gas with all molecules in one half of the container is much less probable than the macrostate with the molecules equally distributed over the whole container (like the state with all coins landed head up is much less probable than the state with half of the coins landed head up and the other half tail up). For this reason, if the initial state has all molecules in one half of the container, then in the course of evolution, the system will come to the most probable state with the molecules equally distributed over the whole container and will stay in this state forever.

We have seen in thermodynamics that an isolated system, initially in a nonequilibrium state, evolves to the equilibrium state characterized by the maximal entropy. This way, one comes to the idea that entropy S and thermodynamic probability w should be related, one being a monotonic function of the other. The form of this function can be found if one notices that entropy is additive, while the thermodynamic probability is multiplicative. If a system consists of two subsystems that weakly interact with each other (which is almost always the case as the intersubsystem interaction is limited to the small region near the interface between them), then $S = S_1 + S_2$ and $w = w_1 w_2$. If one chooses (Boltzmann)

$$S = k_B \ln w, \tag{3.5.1}$$

then $S = S_1 + S_2$ and $w = w_1 w_2$ are in accordance since $\ln(w_1 w_2) = \ln w_1 + \ln w_2$. In Eq. (3.5.1), k_B is the Boltzmann constant,

$$k_B = 1.38 \times 10^{-23} \text{ J K}^{-1}. \tag{3.5.2}$$

It will be shown that the statistically defined entropy above coincides with the thermodynamic entropy at equilibrium. On the other hand, statistical entropy is well defined for nonequilibrium states as well, whereas the thermodynamic entropy is usually undefined off equilibrium.

3.6 Boltzmann distribution and connection with thermodynamics

In this section, we obtain the distribution of particles over energy levels i as the most probable macrostate by maximizing its thermodynamic probability w. We label quantum states by the index i and use Eq. (3.4.1). The task is to find the maximum of w with respect to all N_i that satisfy the constraints (3.1.2) and (3.1.3). Practically, it is more convenient to maximize $\ln w$ than w itself. Using the method of Lagrange multipliers, one searches for the maximum of the target function

$$\Phi(N_1, N_2, \cdots, N_n) = \ln w + \alpha \sum_i N_i - \beta \sum_i \varepsilon_i N_i, \qquad (3.6.1)$$

where α and β are Lagrange multipliers with (arbitrary) signs chosen anticipating the final result. The maximum satisfies

$$\frac{\partial \Phi}{\partial N_i} = 0, \quad i = 1, 2, \dots. \qquad (3.6.2)$$

As we are interested in the behavior of macroscopic systems with $N_i \gg 1$, the factorials can be simplified with the help of Eq. (3.3.1) that takes the form

$$\ln N! \cong N \ln N - N + \ln \sqrt{2\pi N}. \qquad (3.6.3)$$

Neglecting the relatively small last term in this expression, one obtains

$$\Phi \cong \ln N! - \sum_j N_j \ln N_j + \sum_j N_j + \alpha \sum_j N_j - \beta \sum_j \varepsilon_j N_j. \qquad (3.6.4)$$

The first term here is a constant, and it won't contribute to the derivatives $\partial\Phi/\partial N_i$. In the latter, the only contribution comes from the terms with $j = i$ in the above expression. One obtains the equations

$$\frac{\partial\Phi}{\partial N_i} = -\ln N_i + \alpha - \beta\varepsilon_i = 0 \qquad (3.6.5)$$

that yield

$$N_i = e^{\alpha - \beta\varepsilon_i}, \qquad (3.6.6)$$

the Boltzmann distribution with yet undefined Lagrange multipliers α and β. The latter can be found from Eqs. (3.1.2) and (3.1.3) in terms of N and U. Summing Eq. (3.6.6) over i, one obtains

$$N = e^{\alpha}Z, \quad \alpha = \ln(N/Z), \qquad (3.6.7)$$

where

$$Z = \sum_i e^{-\beta\varepsilon_i} \qquad (3.6.8)$$

is the so-called *partition function* (German *Zustandssumme*) that plays a major role in statistical physics. Then, eliminating α from Eq. (3.6.6) yields

$$N_i = \frac{N}{Z}e^{-\beta\varepsilon_i}, \qquad (3.6.9)$$

the Boltzmann distribution. After that, for the internal energy U, one obtains

$$U = \sum_i \varepsilon_i N_i = \frac{N}{Z}\sum_i \varepsilon_i e^{-\beta\varepsilon_i} = -\frac{N}{Z}\frac{\partial Z}{\partial\beta} \qquad (3.6.10)$$

or

$$U = -N\frac{\partial\ln Z}{\partial\beta}. \qquad (3.6.11)$$

This formula implicitly defines the Lagrange multiplier β as a function of U.

The statistical entropy, Eq. (3.5.1), within the Stirling approximation becomes

$$S = k_B \ln w = k_B \left(N \ln N - \sum_i N_i \ln N_i \right). \qquad (3.6.12)$$

Inserting here Eq. (3.6.6) and α from Eq. (3.6.7), one obtains

$$\frac{S}{k_B} = N \ln N - \sum_i N_i \left(\alpha - \beta \varepsilon_i \right) = N \ln N - \alpha N + \beta U = N \ln Z + \beta U.$$
$$(3.6.13)$$

The statistical entropy depends only on the parameter β. Its differential is given by

$$dS = \frac{dS}{d\beta} d\beta = \left(N \frac{\partial \ln Z}{\partial \beta} + U + \beta \frac{dU}{d\beta} \right) d\beta = \beta dU. \qquad (3.6.14)$$

Comparing this with the thermodynamic relation $dU = TdS$ at $V = 0$, one identifies

$$\beta = \frac{1}{k_B T}. \qquad (3.6.15)$$

Now, with the help of $dT/d\beta = -k_B T^2$, one can represent the internal energy, Eq. (3.6.11), via the derivative with respect to T as

$$U = N k_B T^2 \frac{\partial \ln Z}{\partial T}. \qquad (3.6.16)$$

Equation (3.6.13) becomes

$$S = N k_B \ln Z + \frac{U}{T}. \qquad (3.6.17)$$

From here, one obtains the statistical formula for the free energy

$$F = U - TS = -N k_B T \ln Z. \qquad (3.6.18)$$

One can see that the partition function Z contains the complete information of the system's thermodynamics since other quantities such as pressure P follow from F. In particular, one can check $\partial F / \partial T = -S$.

3.7   Quantum states and energy levels

3.7.1   Stationary Schrödinger equation

In the formalism of quantum mechanics, quantized states and their energies E are the solutions of the *eigenvalue* problem for a matrix or for a differential operator. In the latter case, the problem is formulated as the so-called stationary Schrödinger equation

$$\hat{H}\Psi = E\Psi, \qquad (3.7.1)$$

where $\Psi = \Psi(\mathbf{r})$ is the complex function called *wave function*. The physical interpretation of the wave function is that $|\Psi(\mathbf{r})|^2$ gives the probability for a particle to be found near the space point $\mathbf{r}$. As above, the number of measurements dN of the total N measurements in which the particle is found in the elementary volume $d^3r = dxdydz$ around $\mathbf{r}$ is given by

$$dN = N\,|\Psi(\mathbf{r})|^2\,d^3r. \qquad (3.7.2)$$

The wave function satisfies the normalization condition

$$1 = \int d^3r\,|\Psi(\mathbf{r})|^2. \qquad (3.7.3)$$

The operator $\hat{H}$ in Eq. (3.7.1) is the so-called Hamilton operator or Hamiltonian. For one particle, it is the sum of its kinetic and potential energies

$$\hat{H} = \frac{\hat{\mathbf{p}}^2}{2m} + U(\mathbf{r}), \qquad (3.7.4)$$

where the classical momentum $\mathbf{p}$ is replaced by the operator

$$\hat{\mathbf{p}} = -i\hbar\frac{\partial}{\partial\mathbf{r}}. \qquad (3.7.5)$$

The Schrödinger equation can be formulated both for single particles and for systems of particles. In this course, we will restrict ourselves to single particles. In this case, the notation ε will be used for single-particle energy levels instead of E. One can see that Eq. (3.7.1) is a second-order linear differential equation. It is an ordinary differential equation in one dimension and a partial differential equation in two or more dimensions. If there is a potential energy, this is a linear differential equation with variable coefficients that can be solved analytically only in special cases. In the case, $U = 0$, this is a linear differential equation with constant coefficients that is easy to solve analytically.

An important component of the quantum formalism is the boundary conditions for the wave function. In particular, for a particle inside a box with rigid walls, the boundary condition is $\Psi=0$ at the walls, so that $\Psi(\mathbf{r})$ joins smoothly with the value $\Psi(\mathbf{r}) = 0$ everywhere outside the box. In this case, it is also guaranteed that $|\Psi(\mathbf{r})|^2$ is integrable and Ψ can be normalized according to Eq. (3.7.3). It turns out that the solution of Eq. (3.7.1) that satisfies the boundary conditions exists only for a discrete set of E values that are called *eigenvalues*. The corresponding Ψ are called *eigenfunctions*, and all together are called *eigenstates*. Eigenvalue problems, both for matrices and differential operators, were known in mathematics before the advent of quantum mechanics. The creators of quantum mechanics, mainly Schrödinger and Heisenberg, realized that this mathematical formalism could accurately describe the quantization of energy levels observed in experiments. While Schrödinger formulated his famous Schrödinger equation, Heisenberg made a major contribution to the description of quantum systems with matrices.

3.7.2   Energy levels of a particle in a box

As an illustration, consider a particle in a 1D rigid box, $0 \le x \le L$. In this case, the momentum becomes

$$\hat{p} = -i\hbar \frac{d}{dx}, \tag{3.7.6}$$

and Eq. (3.7.1) takes the form

$$-\frac{\hbar^2}{2m}\frac{d^2}{dx^2}\Psi(x) = \varepsilon\Psi(x), \tag{3.7.7}$$

which can be represented as

$$\frac{d^2}{dx^2}\Psi(x) + k^2\Psi(x) = 0, \quad k^2 \equiv \frac{2m\varepsilon}{\hbar^2}. \tag{3.7.8}$$

The solution of this equation satisfying the boundary conditions $\Psi(0) = \Psi(L) = 0$ has the form

$$\Psi_\nu(x) = A\sin{(k_\nu x)}, \quad k_\nu = \frac{\pi}{L}\nu, \quad \nu = 1, 2, 3, \ldots, \tag{3.7.9}$$

where eigenstates are labeled by the index ν. The constant A following from Eq. (3.7.3) is $A = \sqrt{2/L}$. The energy eigenvalues are given by

$$\varepsilon_\nu = \frac{\hbar^2 k_\nu^2}{2m} = \frac{\pi^2 \hbar^2 \nu^2}{2mL^2}. \tag{3.7.10}$$

One can see the energy ε is quadratic in the momentum $p = \hbar k$ (de Broglie relation), as it should be, but the energy levels are discrete because of the quantization. For a large system size L or a large mass m, the energy difference between the neighboring levels, $\varepsilon_{\nu+1} - \varepsilon_\nu$, becomes small, thus the energy levels become quasicontinuous. The lowest-energy level with $\nu = 1$ is called the *ground state*.

For a 3D box with sides L_x, L_y, and L_z, one has to solve the Schrödinger equation

$$-\frac{\hbar^2}{2m}\left(\frac{d^2}{dx^2} + \frac{d^2}{dy^2} + \frac{d^2}{dz^2}\right)\Psi(x) = \varepsilon\Psi(x), \tag{3.7.11}$$

with similar boundary conditions. The solution factorizes and has the form

$$\Psi_{\nu_x,\nu_y,\nu_z}(x, y, z) = A\sin{(k_{\nu_x}x)}\sin{(k_{\nu_y}x)}\sin{(k_{\nu_z}x)},$$

$$k_\alpha \equiv k_{\nu_\alpha} = \frac{\pi}{L_\alpha}\nu_\alpha, \tag{3.7.12}$$

where $\nu_\alpha = 1, 2, 3, \ldots$ and $\alpha = x, y, z$. The energy levels are

$$\varepsilon = \frac{\hbar^2 \mathbf{k}^2}{2m} = \frac{\hbar^2 \left(k_x^2 + k_y^2 + k_z^2\right)}{2m} \tag{3.7.13}$$

and parametrized by the three quantum numbers ν_α. The ground state is $(\nu_x, \nu_y, \nu_z) = (1, 1, 1)$. One can order the states in increasing ε and number them by the index j, the ground state being $j = 1$. If $L_x = L_y = L_z = L$, then

$$\varepsilon_{\nu_x,\nu_y,\nu_z} = \frac{\pi^2\hbar^2}{2mL^2}\left(\nu_x^2 + \nu_y^2 + \nu_z^2\right), \tag{3.7.14}$$

and the same value of ε_j can be realized for different sets of ν_x, ν_y, and ν_z, e.g., (1,5,12) and (5,12,1). In this case, it is convenient to ascribe the same index j to all these states. The number of different sets of (ν_x, ν_y, ν_z) having the same ε_j is called degeneracy and is denoted as g_j. States with $\nu_x = \nu_y = \nu_z$ have $g_j = 1$, and they are called nondegenerate. If only two of the numbers ν_x, ν_y, and ν_z coincide, the degeneracy is $g_j = 3$. If all numbers are different, $g_j = 3! = 6$. If one sums over the energy levels parametrized by j, one has to multiply the summands by the corresponding degeneracies.

3.7.3 Density of states and classical limit of statistical physics

For systems of a large size or particles of a large mass, energy levels become so finely quantized that the energy spectrum can be considered as continuous. In this case, one can define the density of states $\rho(\varepsilon)$ as the number of energy levels dn in the interval $d\varepsilon$, i.e.,

$$dn = \rho(\varepsilon)d\varepsilon. \tag{3.7.15}$$

It is convenient to start the calculation of $\rho(\varepsilon)$ by introducing the number of states dn in the "elementary volume" $d\nu_x d\nu_y d\nu_z$, considering ν_x, ν_y, and ν_z as continuous. The result obviously is

$$dn = d\nu_x d\nu_y d\nu_z, \tag{3.7.16}$$

i.e., the corresponding density of states is 1. Now, one can rewrite the same number of states in terms of the wave vector $\mathbf{k}$ using Eq. (3.7.13) as

$$dn = \frac{V}{\pi^3} dk_x dk_y dk_z, \tag{3.7.17}$$

where $V = L_x L_y L_z$ is the volume of the box. After that, one can go over to the number of states within the shell dk, as was done for the distribution function of the molecules over velocities. Taking into account that k_x, k_y, and k_z are all positive, this shell is not a complete spherical shell but 1/8 of it. Thus,

$$dn = \frac{V}{\pi^3} \frac{4\pi}{8} k^2 dk. \tag{3.7.18}$$

The last step is to change the variable from k to ε using Eq. (3.7.13). With

$$k^2 = \frac{2m}{\hbar^2}\varepsilon, \quad k = \sqrt{\frac{2m}{\hbar^2}}\sqrt{\varepsilon}, \quad \frac{dk}{d\varepsilon} = \frac{1}{2}\sqrt{\frac{2m}{\hbar^2}}\frac{1}{\sqrt{\varepsilon}}, \quad (3.7.19)$$

one obtains Eq. (3.7.15) with

$$\rho(\varepsilon) = \frac{V}{(2\pi)^2}\left(\frac{2m}{\hbar^2}\right)^{3/2}\sqrt{\varepsilon}. \quad (3.7.20)$$

For a quasicontinuous energy spectrum, one can replace summation over the energy levels i by integration over the energy ε

$$\sum_i \cdots \Rightarrow \int d\varepsilon\,\rho(\varepsilon)\cdots \quad (3.7.21)$$

In statistical physics, sums with the Boltzmann factor $e^{-\beta\varepsilon_i}$ arise. The condition for replacing summation by integration is that the Boltzmann exponentials for the neighboring energy levels are close to each other, i.e., $e^{-\beta\varepsilon_i} - e^{-\beta\varepsilon_{i+1}} \ll e^{-\beta\varepsilon_i}$. This requires finely spaced levels and/or temperatures high enough: $\beta\Delta\varepsilon \ll 1$ or

$$\Delta\varepsilon \ll k_B T, \quad (3.7.22)$$

where $\Delta\varepsilon \equiv \varepsilon_{i+1} - \varepsilon_i$. In particular, the partition function, Eq. (3.6.8), as an integral has the form

$$Z = \int d\varepsilon\,\rho(\varepsilon)\,e^{-\beta\varepsilon}. \quad (3.7.23)$$

The number of particles dN_ε in quantum states within the energy interval $d\varepsilon$ is the product of the number of particles in one state $N_i \Rightarrow N(\varepsilon)$ and the number of states dn_ε in this energy interval. Using Boltzmann distribution, Eq. (3.6.9) for the former and Eq. (3.7.15) for the latter, one obtains the Boltzmann distribution over the energies

$$dN_\varepsilon = N(\varepsilon)dn_\varepsilon = \frac{N}{Z}e^{-\beta\varepsilon}\rho(\varepsilon)\,d\varepsilon. \quad (3.7.24)$$

There is no information about quantum energy levels ε_i in the formula for dN_ε. Planck's constant that usually enters $\rho(\varepsilon)$, as in Eq. (3.7.20), cancels because it enters the numerator and the denominator in the same way. As a result, one obtains essentially classical results for physical quantities.

3.8 Statistical thermodynamics of the ideal gas

In this section, we demonstrate how the results for the ideal gas, previously obtained within the molecular theory, follow from the more general framework of statistical physics formulated in the preceding section. Consider an ideal gas in a sufficiently large 3D container. In this case, the energy levels of the system, see Eq. (3.7.13), are quantized so finely that one can introduce the density of states $\rho(\varepsilon)$ defined by Eq. (3.7.15). For finely quantized levels, one can replace summation by integration and use Eq. (3.7.23) for the partition function. This yields

$$Z = \frac{V}{(2\pi)^2}\left(\frac{2m}{\hbar^2}\right)^{3/2}\int_0^\infty d\varepsilon\sqrt{\varepsilon}e^{-\beta\varepsilon}$$

$$= \frac{V}{(2\pi)^2}\left(\frac{2m}{\hbar^2}\right)^{3/2}\frac{\sqrt{\pi}}{2\beta^{3/2}} = V\left(\frac{mk_BT}{2\pi\hbar^2}\right)^{3/2}. \quad (3.8.1)$$

Now, Boltzmann distribution over the energies, Eq. (3.7.24), becomes

$$dN_\varepsilon = N\frac{(2\pi)^2}{V}\left(\frac{\hbar^2}{2m}\right)^{3/2}\frac{2\beta^{3/2}}{\sqrt{\pi}}\frac{V}{(2\pi)^2}\left(\frac{2m}{\hbar^2}\right)^{3/2}e^{-\beta\varepsilon}\sqrt{\varepsilon}d\varepsilon$$

$$= N\frac{2\beta^{3/2}}{\sqrt{\pi}}e^{-\beta\varepsilon}\sqrt{\varepsilon}d\varepsilon. \quad (3.8.2)$$

Planck's constant $\hbar$ that links to quantum mechanics disappeared from the final result. Using the classical relation $\varepsilon = mv^2/2$, and thus, $d\varepsilon = mvdv$, one can obtain the formula for the number of particles in the speed interval dv

$$dN_v = Nf(v)dv, \quad (3.8.3)$$

where the speed distribution function $f(v)$ is given by

$$f(v) = \frac{2\beta^{3/2}}{\sqrt{\pi}}e^{-\beta\varepsilon}\sqrt{\varepsilon}mv = \left(\frac{m}{2\pi k_BT}\right)^{3/2}4\pi v^2\exp\left(-\frac{mv^2}{2k_BT}\right). \quad (3.8.4)$$

This result coincides with the Maxwell distribution function obtained earlier from the molecular theory of gases.

The internal energy of the ideal gas is its kinetic energy

$$U = N\bar{\varepsilon}, \tag{3.8.5}$$

$\bar{\varepsilon} = m\bar{v}^2/2$ being the average kinetic energy of an atom. The latter can be calculated with the help of the energy distribution function or the speed distribution function above, as was done in the molecular theory of gases. The result has the form

$$\bar{\varepsilon} = \frac{3}{2}k_B T. \tag{3.8.6}$$

The same result can be obtained from Eq. (3.6.11):

$$U = -N\frac{\partial \ln Z}{\partial \beta} = -N\frac{\partial}{\partial \beta} \ln\left[V\left(\frac{m}{2\pi\hbar^2\beta}\right)^{3/2}\right]$$

$$= \frac{3}{2}N\frac{\partial}{\partial \beta}\ln\beta = \frac{3}{2}N\frac{1}{\beta} = \frac{3}{2}Nk_B T. \tag{3.8.7}$$

After that, the known result for the heat capacity $C_V = (\partial U/\partial T)_V$ follows.

The pressure P is defined from the free energy $F = -Nk_B T \ln Z$ by the thermodynamic formula

$$P = -\left(\frac{\partial F}{\partial V}\right)_T. \tag{3.8.8}$$

With the help of Eq. (3.8.1), one obtains

$$P = Nk_B T\frac{\partial \ln Z}{\partial V} = Nk_B T\frac{\partial \ln V}{\partial V} = \frac{Nk_B T}{V}, \tag{3.8.9}$$

which amounts to the equation of state of the ideal gas, $PV = Nk_B T$.

Let us calculate the entropy of the ideal gas in the box. Using either $S = -(\partial F/\partial T)_V$ or simply Eq. (3.6.17), one obtains

$$S = Nk_B \ln Z + \frac{3}{2}Nk_B = Nk_B \ln\left[V\left(\frac{mk_B T}{2\pi\hbar^2}\right)^{3/2}\right]$$

$$= \frac{3}{2}Nk_B \ln\left[V^{2/3}\frac{mk_B T}{2\pi\hbar^2}\right]. \tag{3.8.10}$$

This result coincides with the thermodynamic formula for the entropy of the ideal gas,

$$S = C_V \ln \left(TV^{\gamma-1} \right) + S_0. \qquad (3.8.11)$$

In our case of three degrees of freedom, $f = 3$, one has $C_V = (3/2)\, Nk_B$ and $\gamma - 1 = 2/f = 2/3$. In addition, statistical physics captures the constant term in the entropy

$$S_0 = \frac{3}{2} Nk_B \ln \frac{mk_B}{2\pi\hbar^2}, \qquad (3.8.12)$$

which depends on Planck's constant $\hbar$ and thus has a quantum origin.

The entropy found above does not satisfy the third law of thermodynamics as it does not vanish in the limit $T \to 0$. The reason is that we replaced summation over quantum levels by integration, which fails at very low temperatures as Eq. (3.7.22) is not satisfied. If the discreteness of the levels at these very low temperatures is taken into account, one can see that all particles fall into the ground state, while the number of particles in excited states is exponentially small and can be neglected. If the ground state is nondegenerate, the partition function according to Eq. (3.6.8) becomes $e^{-\beta\varepsilon_0}$, where ε_0 is the ground-state energy. Then Eq. (3.6.17) yields

$$S = Nk_B \ln Z + \frac{U}{T} = -Nk_B \frac{\varepsilon_0}{k_B T} + \frac{N\varepsilon_0}{T} = 0, \qquad (3.8.13)$$

as it should be.

3.9 Statistical thermodynamics of harmonic oscillators

Consider an ensemble of N identical harmonic oscillators, each of them described by the Hamiltonian

$$\hat{H} = \frac{\hat{p}^2}{2m} + \frac{kx^2}{2}, \qquad (3.9.1)$$

where the momentum operator $\hat{p}$ is given by Eq. (3.7.6) and k is the spring constant. This theoretical model can describe, for instance, vibrational degrees of freedom of diatomic molecules. In this case, x is the elongation of the chemical bond between the two atoms, relative to the equilibrium bond length. The stationary Schrödinger equation (3.7.1) for a harmonic oscillator becomes

$$\left(-\frac{\hbar^2}{2m}\frac{d^2}{dx^2} + \frac{kx^2}{2}\right)\Psi(x) = \varepsilon\Psi(x). \tag{3.9.2}$$

The boundary conditions for this equation require that $\Psi(\pm\infty) = 0$ and the integral in Eq. (3.7.3) converges. In contrast to Eq. (3.7.7), this is a linear differential equation with a *variable* coefficient. Such differential equations in general do not have solutions in terms of known functions. In some cases, the solution can be expressed through *special functions*, such as hypergeometrical and Bessel functions. The solution of Eq. (3.9.2) can be found, but this task belongs to quantum mechanics courses. The main result that we need here is that the energy eigenvalues of the harmonic oscillator have the form

$$\varepsilon_\nu = \hbar\omega_0\left(\nu + \frac{1}{2}\right), \quad \nu = 0, 1, 2, \ldots, \tag{3.9.3}$$

where $\omega_0 = \sqrt{k/m}$ is the frequency of oscillations. The energy level with $\nu = 0$ is the ground state. The ground-state energy $\varepsilon_0 = \hbar\omega_0/2$ is not zero, as would be the case for a classical oscillator. This quantum ground-state energy is called *zero-point energy*. It is irrelevant in the calculation of the heat capacity of an ensemble of harmonic oscillators.

The partition function of a harmonic oscillator is

$$Z = \sum_{\nu=0}^{\infty} e^{-\beta\varepsilon_\nu} = e^{-\beta\hbar\omega_0/2}\sum_{\nu=0}^{\infty}\left(e^{-\beta\hbar\omega_0}\right)^{\nu}$$

$$= \frac{e^{-\beta\hbar\omega_0/2}}{1 - e^{-\beta\hbar\omega_0}} = \frac{2}{\sinh\left(\beta\hbar\omega_0/2\right)}, \tag{3.9.4}$$

where the result for the geometrical progression

$$1 + x + x^2 + x^3 + \cdots = (1 - x)^{-1}, \quad x < 1 \tag{3.9.5}$$

was used. The hyperbolic functions are defined by

$$\sinh(x) \equiv \frac{e^x - e^{-x}}{2}, \quad \cosh(x) \equiv \frac{e^x + e^{-x}}{2},$$

$$\tanh(x) \equiv \frac{\sinh(x)}{\cosh(x)}, \quad \coth(x) \equiv \frac{\cosh(x)}{\sinh(x)} = \frac{1}{\tanh(x)}. \tag{3.9.6}$$

We will be using the formulas

$$\sinh(x)' = \cosh(x), \quad \cosh(x)' = \sinh(x), \tag{3.9.7}$$

and

$$\sinh(x) \cong \begin{cases} x, & x \ll 1, \\ e^x/2, & x \gg 1, \end{cases} \quad \tanh(x) \cong \begin{cases} x, & x \ll 1, \\ 1, & x \gg 1. \end{cases} \tag{3.9.8}$$

The internal (average) energy of the ensemble of oscillators is given by Eq. (3.6.11). This yields

$$U = -N\frac{\partial \ln Z}{\partial \beta} = N\frac{\partial \ln \sinh (\beta\hbar\omega_0/2)}{\partial \beta} = N\frac{\hbar\omega_0}{2} \coth \left(\frac{\beta\hbar\omega_0}{2}\right) \tag{3.9.9}$$

or

$$U = N\frac{\hbar\omega_0}{2} \coth \left(\frac{\hbar\omega_0}{2k_BT}\right). \tag{3.9.10}$$

Another way of writing the internal energy is

$$U = N\hbar\omega_0 \left(\frac{1}{e^{\beta\hbar\omega_0} - 1} + \frac{1}{2}\right), \tag{3.9.11}$$

where the constant term with $1/2$ is the zero-point energy. The limiting low- and high-temperature cases of this formula are

$$U \cong \begin{cases} N\hbar\omega_0/2, & k_BT \ll \hbar\omega_0, \\ Nk_BT, & k_BT \gg \hbar\omega_0. \end{cases} \tag{3.9.12}$$

In the low temperature limit, almost all oscillators are in their ground states since $e^{-\beta\hbar\omega_0} \ll 1$. Thus, the main term that contributes to the partition function in Eq. (3.9.4) is $\nu = 0$. Correspondingly, U is the sum of the ground-state energies of all oscillators. At high temperatures, Planck's constant $\hbar$ disappears, and the result becomes classical. In this case, $e^{-\beta\hbar\omega_0}$ is only slightly smaller than one, so that very many different n contribute to Z in Eq. (3.9.4). Thus, one can replace summation by integration, as was done above for the particle in a potential box, to obtain the second line of Eq. (3.9.12).

One can see that the crossover between the two limiting cases corresponds to $k_BT \sim \hbar\omega_0$. In the high-temperature limit $k_BT \gg \hbar\omega_0$, many low-lying energy levels are populated. The top populated level ν^* can be estimated from $k_B T \sim \hbar\omega_0 \nu^*$, so that

$$\nu^* \sim \frac{k_B T}{\hbar\omega_0} \gg 1. \qquad (3.9.13)$$

The probability of finding an oscillator in the states with $\nu \gg \nu^*$ is exponentially small.

The heat capacity is defined by

$$C = \frac{dU}{dT} = N\frac{\hbar\omega_0}{2}\left(-\frac{1}{\sinh^2\left[\hbar\omega_0/(2k_B T)\right]}\right)\left(-\frac{\hbar\omega_0}{2k_B T^2}\right)$$

$$= Nk_B\left(\frac{\hbar\omega_0/(2k_B T)}{\sinh\left[\hbar\omega_0/(2k_B T)\right]}\right)^2. \qquad (3.9.14)$$

This formula has limiting cases

$$C \cong \begin{cases} Nk_B\left(\dfrac{\hbar\omega_0}{k_B T}\right)^2 \exp\left(-\dfrac{\hbar\omega_0}{k_B T}\right), & k_B T \ll \hbar\omega_0, \\[2ex] Nk_B, & k_B T \gg \hbar\omega_0. \end{cases} \qquad (3.9.15)$$

One can see that at high temperatures, the heat capacity can be written as

$$C = \frac{f}{2}Nk_B, \qquad (3.9.16)$$

where the effective number of degrees of freedom for an oscillator is $f = 2$. The explanation of the additional factor 2 is that the oscillator has not only the kinetic but also the potential energy, and the average values of these two energies are the same. Thus, the total amount of energy in a vibrational degree of freedom doubles with respect to the translational and rotational degrees of freedom.

At low temperatures, the vibrational heat capacity above becomes exponentially small. One says that vibrational degrees of freedom are getting frozen out at low temperatures.

The heat capacity of an ensemble of harmonic oscillators in the whole temperature range, Eq. (3.9.14), is plotted in Fig. 3.2.

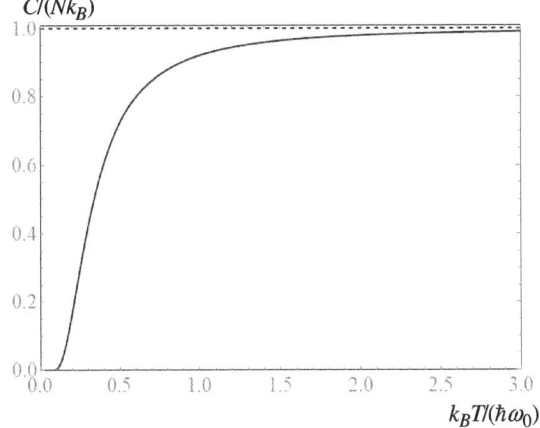

$C/(Nk_B)$

$k_B T/(\hbar\omega_0)$

Figure 3.2. Heat capacity of harmonic oscillators.

The average quantum number of an oscillator is given by

$$n \equiv \langle \nu \rangle = \frac{1}{Z} \sum_{\nu=0}^{\infty} \nu e^{-\beta \varepsilon_\nu}. \tag{3.9.17}$$

Using Eq. (3.9.3), one can calculate this sum as follows:

$$
\begin{aligned}
n &= \frac{1}{Z} \sum_{\nu=0}^{\infty} \left(\frac{1}{2} + \nu \right) e^{-\beta \varepsilon_\nu} - \frac{1}{Z} \sum_{\nu=0}^{\infty} \frac{1}{2} e^{-\beta \varepsilon_\nu} \\
&= \frac{1}{\hbar\omega_0} \frac{1}{Z} \sum_{\nu=0}^{\infty} \varepsilon_\nu e^{-\beta \varepsilon_\nu} - \frac{Z}{2Z} \\
&= -\frac{1}{\hbar\omega_0} \frac{1}{Z} \frac{\partial Z}{\partial \beta} - \frac{1}{2} = \frac{U}{N\hbar\omega_0} - \frac{1}{2}.
\end{aligned} \tag{3.9.18}
$$

Finally, with the help of Eq. (3.9.11), one finds

$$n = \frac{1}{e^{\beta\hbar\omega_0} - 1} = \frac{1}{\exp\left[\hbar\omega_0/(k_B T)\right] - 1}. \tag{3.9.19}$$

This is the Bose–Einstein distribution that will be considered later. The meaning of it is the following. Considering the oscillator as a "box" or "mode", one can ask what the average number of quanta (i.e., "particles" or quasiparticles) is in this box at a given temperature. The latter is given by the formula above.

Substituting Eq. (3.9.19) into Eq. (3.9.11), one obtains the nice formula

$$U = N\hbar\omega_0 \left(n + \frac{1}{2} \right), \tag{3.9.20}$$

resembling Eq. (3.9.3). At low temperatures, $k_B T \ll \hbar\omega_0$, the average quantum number n becomes exponentially small. This means that the oscillator is predominantly in its ground state, $\nu = 0$. At high temperatures, $k_B T \gg \hbar\omega_0$, Eq. (3.9.19) yields

$$n \cong \frac{k_B T}{\hbar\omega_0}, \tag{3.9.21}$$

which has the same order of magnitude as the top populated quantum level number ν^* given by Eq. (3.9.13).

The density of states $\rho(\varepsilon)$ for the harmonic oscillator can be easily found. The number of levels dn_ν in the interval $d\nu$ is

$$dn_\nu = d\nu \tag{3.9.22}$$

(i.e., there is only one level in the interval $d\nu = 1$). Then, changing the variables with the help of Eq. (3.9.3), one finds the number of levels dn_ε in the interval $d\varepsilon$ as

$$dn_\varepsilon = \frac{d\nu}{d\varepsilon} d\varepsilon = \left(\frac{d\varepsilon}{d\nu} \right)^{-1} d\varepsilon = \rho(\varepsilon) \, d\varepsilon, \tag{3.9.23}$$

where

$$\rho(\varepsilon) = \frac{1}{\hbar\omega_0}. \tag{3.9.24}$$

To conclude this section, let us reproduce the classical high-temperature results by a simpler method. At high temperatures, many levels are populated, and the distribution function $e^{-\beta\varepsilon_\nu}$ only slightly changes from one level to the other. Indeed, its relative change is

$$\frac{e^{-\beta\varepsilon_\nu} - e^{-\beta\varepsilon_{\nu+1}}}{e^{-\beta\varepsilon_\nu}} = 1 - e^{-\beta(\varepsilon_{\nu+1} - \varepsilon_\nu)} = 1 - e^{-\beta\hbar\omega_0}$$

$$\cong 1 - 1 + \beta\hbar\omega_0 = \frac{\hbar\omega_0}{k_B T} \ll 1. \tag{3.9.25}$$

In this situation, one can replace summation in Eq. (3.9.4) by integration over ε using the density of states of Eq. (3.9.24):

$$Z = \int_0^\infty d\varepsilon \rho\left(\varepsilon\right) e^{-\beta\varepsilon} = \frac{1}{\hbar\omega_0} \int_0^\infty d\varepsilon e^{-\beta\varepsilon} = \frac{1}{\hbar\omega_0 \beta} = \frac{k_B T}{\hbar\omega_0}.$$

$$(3.9.26)$$

Also, one can integrate over ν, considering it as continuous and not using the density of states explicitly:

$$Z = \int_0^\infty d\nu e^{-\beta\varepsilon} = \int_0^\infty d\varepsilon \frac{d\nu}{d\varepsilon} e^{-\beta\varepsilon} = \frac{1}{\hbar\omega_0} \int_0^\infty d\varepsilon e^{-\beta\varepsilon}$$

$$= \frac{1}{\hbar\omega_0 \beta} = \frac{k_B T}{\hbar\omega_0}.$$

$$(3.9.27)$$

The result of this calculation is in accordance with Eq. (3.9.24). Now, the internal energy U is given by

$$U = -N\frac{\partial \ln Z}{\partial \beta} = -N\frac{\partial \ln \frac{1}{\beta} + \cdots}{\partial \beta} = N\frac{\partial \ln \beta + \cdots}{\partial \beta} = \frac{N}{\beta} = Nk_B T,$$

$$(3.9.28)$$

which coincides with the second line of Eq. (3.9.12).

In 1907, Albert Einstein proposed a theory of thermal properties of solids based on the assumption of the existence therein of independent thermally excited harmonic oscillators. The essential moment was that Einstein applied Planck's idea of quantization, extending it from the electromagnetic radiation (see Section 3.21) to the vibrations of atoms and molecules in solids. Before Einstein's work, it was a puzzle as to why the heat capacity is not constant, as predicted by the classical theory (our high-temperature limit), but becomes small at low temperatures. Figure 3.2 shows the result of Einstein's theory. This was historically the third application of the idea of quantization after the seminal 1900 work of Planck and Einstein's theory of the photoeffect of 1905. All these three achievements were accomplished before quantum mechanics was finalized about 1927. Einstein's result at low temperatures is only qualitative as the concept ignores the interaction of the oscillations of different atoms. The accurate theory by Debye (see Section 3.11), leads to the dependence $C_V \propto T^3$ in 3D solids at low temperatures. Still, Figs. 3.2 and 3.4 look pretty similar.

3.10   Statistical thermodynamics of rotators

The rotational kinetic energy of a rigid body, expressed with respect to its *principal axes*, reads

$$E_{\text{rot}} = \frac{1}{2} I_1 \omega_1^2 + \frac{1}{2} I_2 \omega_2^2 + \frac{1}{2} I_3 \omega_3^2. \tag{3.10.1}$$

We will restrict our consideration to the axially symmetric body with $I_1 = I_2 = I$ since for the fully asymmetric body, the quantum energy levels cannot be found analytically. In this case, E_{rot} can be rewritten in terms of the angular momentum components $L_\alpha = I_\alpha \omega_\alpha$, where $\alpha = 1, 2, 3$, as follows:

$$E_{\text{rot}} = \frac{L_1^2}{2I_1} + \frac{L_2^2}{2I_2} + \frac{L_3^2}{2I_3} = \frac{\mathbf{L}^2}{2I} + \frac{1}{2} \left(\frac{1}{I_3} - \frac{1}{I} \right) L_3^2, \tag{3.10.2}$$

where $\mathbf{L}^2 = L_1^2 + L_2^2 + L_3^2$.

In the important case of a diatomic molecule with two identical atoms, the center of mass is located between the two atoms at the distance $d/2$ from each, where d is the distance between the atoms. The moment of inertia of the molecule with respect to the three axis going through both atoms is zero, $I_3 = 0$. The moments of inertia with respect to the 1 and 2 axes are equal to each other:

$$I_1 = I_2 = I = 2 \times M \left(\frac{d}{2} \right)^2 = \frac{1}{2} M d^2, \tag{3.10.3}$$

where M is the mass of the atom. In our case, $I_3 = 0$, there is no kinetic energy associated with rotation around the three axis. Then Eq. (3.10.2) requires $L_3 = 0$, i.e., the angular momentum of a diatomic molecule is perpendicular to its axis.

In quantum mechanics, $\mathbf{L}$ and L_3 become operators, and the rotational energy above becomes the Hamiltonian, $E_{\text{rot}} \Rightarrow \hat{H}_{\text{rot}}$. The solution of the stationary Schrödinger equation yields eigenvalues of $\mathbf{L}^2$, L_3 in terms of three quantum numbers l, m, and n:

$$\hat{\mathbf{L}}^2 = \hbar^2 l(l+1), \quad l = 0, 1, 2, \ldots \tag{3.10.4}$$

$$\hat{L}_3 = \hbar n, \quad m, n = -l, -l+1, \ldots, l-1, l. \tag{3.10.5}$$

Thus, the energy eigenvalues are given by

$$\varepsilon_{l,n} = \frac{\hbar^2 l(l+1)}{2I} + \frac{\hbar^2}{2}\left(\frac{1}{I_3} - \frac{1}{I}\right) n^2, \qquad (3.10.6)$$

which follows by the substitution of Eqs. (3.10.4) and (3.10.5) into Eq. (3.10.2). The quantum number m accounts for the $2l+1$ different orientations of the vector $\mathbf{L}$ in space. As all these orientations are equivalent, the energy does not depend on m. Thus, there are at least $2l+1$ degenerate energy levels. Further, the axis of the molecule can have $2l+1$ different projections on $\hat{\mathbf{L}}$ parametrized by the quantum number n. In the general case, $I_3 \neq I$, these states are nondegenerate. However, they become degenerate for the fully symmetric body, $I_1 = I_2 = I_3 = I$. In this case, the energy eigenvalues become

$$\varepsilon_l = \frac{\hbar^2 l(l+1)}{2I}, \qquad (3.10.7)$$

and the degeneracy of the quantum level l is $g_l = (2l+1)^2$. For the diatomic molecule, $I_3 = 0$, and the only acceptable value of n is $n = 0$. Thus, one obtains Eq. (3.10.7) again, but with the degeneracy, $g_l = 2l+1$.

In the following, we will consider only the two cases: (a) a fully symmetric body and (b) a diatomic molecule. The rotational partition function for both of them is given by

$$Z = \sum_l g_l e^{-\beta \varepsilon_l}, \quad g_l = (2l+1)^\xi, \quad \xi = \begin{cases} 2, & \text{symmetric body}, \\ 1, & \text{diatomic molecule}. \end{cases}$$
$$(3.10.8)$$

In both cases, Z cannot be calculated analytically in general. In the general axially symmetric case, $I_1 = I_2 \neq I_3$ with $I_3 \neq 0$, one has to perform a numerical summation over both l and n. However, even the sum over l above cannot be calculated analytically, so we will consider the limits of low and high temperatures.

In the low temperature limit, most of the rotators are in their ground state $l = 0$, and very few are in the first excited state $l = 1$. Discarding all other values of l, one obtains

$$Z \cong 1 + 3^\xi \exp\left(-\frac{\beta \hbar^2}{I}\right) \qquad (3.10.9)$$

at low temperatures. The rotational internal energy is given by

$$U = -N\frac{\partial \ln Z}{\partial \beta} = 3^{\xi}\frac{\hbar^2}{I}\exp\left(-\frac{\beta\hbar^2}{I}\right) = 3^{\xi}\frac{\hbar^2}{I}\exp\left(-\frac{\hbar^2}{Ik_BT}\right),$$

$$(3.10.10)$$

and it is exponentially small. The heat capacity is

$$C_V = \left(\frac{\partial U}{\partial T}\right)_V = 3^{\xi}k_B\left(\frac{\hbar^2}{Ik_BT}\right)^2\exp\left(-\frac{\hbar^2}{Ik_BT}\right), \qquad (3.10.11)$$

and it is also exponentially small.

In the high-temperature limit, very many energy levels are thermally populated, so that in Eq. (3.10.8), one can neglect $1 \ll l$ and replace summation over l by integration. With

$$g_l \cong (2l)^{\xi}, \quad \varepsilon_l \equiv \varepsilon \cong \frac{\hbar^2 l^2}{2I}, \quad \frac{\partial \varepsilon}{\partial l} = \frac{\hbar^2 l}{I}, \quad \frac{\partial l}{\partial \varepsilon} = \frac{I}{\hbar^2}\sqrt{\frac{\hbar^2}{2I\varepsilon}} = \sqrt{\frac{I}{2\hbar^2\varepsilon}},$$

$$(3.10.12)$$

one has

$$\mathcal{Z} \cong \int_0^{\infty} dl\, g_l e^{-\beta\varepsilon_l} \cong 2^{\xi}\int_0^{\infty} dl\, l^{\xi}e^{-\beta\varepsilon_l} = 2^{\xi}\int_0^{\infty} d\varepsilon\, \frac{\partial l}{\partial\varepsilon}l^{\xi}e^{-\beta\varepsilon},$$

$$(3.10.13)$$

and further,

$$Z \cong 2^{\xi}\sqrt{\frac{I}{2\hbar^2}}\int_0^{\infty} d\varepsilon\, \frac{1}{\sqrt{\varepsilon}}\left(\frac{2I\varepsilon}{\hbar^2}\right)^{\xi/2}e^{-\beta\varepsilon}$$

$$= 2^{(3\xi-1)/2}\left(\frac{I}{\hbar^2}\right)^{(\xi+1)/2}\int_0^{\infty} d\varepsilon\, \varepsilon^{(\xi-1)/2}e^{-\beta\varepsilon}$$

$$= 2^{(3\xi-1)/2}\left(\frac{I}{\beta\hbar^2}\right)^{(\xi+1)/2}\int_0^{\infty} dx\, x^{(\xi-1)/2}e^{-x} \propto \beta^{-(\xi+1)/2}.$$

$$(3.10.14)$$

Now, the internal energy is given by

$$U = -N\frac{\partial \ln Z}{\partial \beta} = \frac{\xi+1}{2}N\frac{\partial \ln \beta}{\partial \beta} = \frac{\xi+1}{2}N\frac{1}{\beta} = \frac{\xi+1}{2}Nk_BT,$$

$$(3.10.15)$$

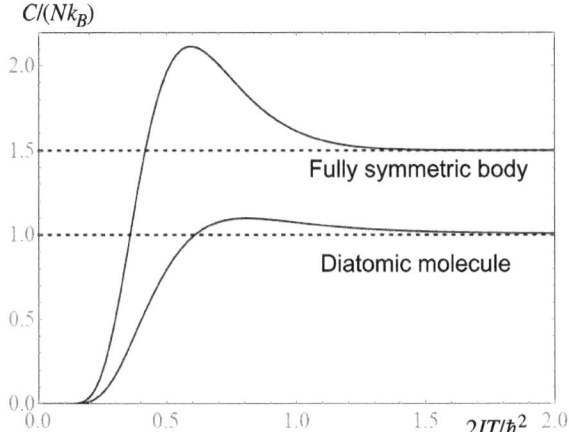

Figure 3.3. Rotational heat capacity of a fully symmetric body and a diatomic molecule.

whereas the heat capacity is

$$C_V = \left(\frac{\partial U}{\partial T}\right)_V = \frac{\xi + 1}{2} N k_B. \qquad (3.10.16)$$

In these two formulas, $f = \xi + 1$ is the number of degrees of freedom, $f = 2$ for a diatomic molecule and $f = 3$ for a fully symmetric rotator. This result confirms the principle of equidistribution of energy over the degrees of freedom in the classical limit, which requires temperatures high enough.

The all-temperature result for the heat capacity of rotators based on the numerical summation in Eq. (3.10.8) is shown in Fig. 3.3. Surprisingly, there is a maximum of the heat capacity in the intermediate temperature region that contrasts the monotonic increase of the heat capacity of the harmonic oscillator shown in Fig. 3.2.

To conclude this section, let us calculate the partition function classically at high temperatures. The calculation can be performed in the general case of all moments of inertia being different. The rotational energy, Eq. (3.10.1), can be written as

$$E_{\text{rot}} = \frac{L_1^2}{2I_1} + \frac{L_2^2}{2I_2} + \frac{L_3^2}{2I_3}. \qquad (3.10.17)$$

The dynamical variables here are the angular momentum components L_1, L_2, and L_3, so that the partition function can be obtained by integration with respect to them:

$$
\begin{aligned}
Z_{\text{class}} &= \int_{-\infty}^{\infty} dL_1 dL_2 dL_3 \exp\left(-\beta E_{\text{rot}}\right) \\
&= \int_{-\infty}^{\infty} dL_1 \exp\left(-\frac{\beta L_1^2}{2I_1}\right) \times \int_{-\infty}^{\infty} dL_2 \exp\left(-\frac{\beta L_2^2}{2I_2}\right) \\
&\quad \times \int_{-\infty}^{\infty} dL_3 \exp\left(-\frac{\beta L_3^2}{2I_3}\right) \\
&= \sqrt{\frac{2I_1}{\beta} \times \frac{2I_2}{\beta} \times \frac{2I_3}{\beta}} \left(\int_{-\infty}^{\infty} dx\, e^{-x^2}\right)^3 \propto \beta^{-3/2}.
\end{aligned} \qquad (3.10.18)
$$

In fact, Z_{class} contains some additional coefficients, for instance, from the integration over the orientations of the molecule. However, these coefficients are irrelevant in the calculation of the internal energy and heat capacity. For the internal energy, one obtains

$$
U = -N\frac{\partial \ln Z}{\partial \beta} = \frac{3}{2}N\frac{\partial \ln \beta}{\partial \beta} = \frac{3}{2}Nk_BT, \qquad (3.10.19)
$$

which coincides with Eq. (3.10.15) with $\xi = 2$. In the case of the diatomic molecule, the energy has the form

$$
E_{\text{rot}} = \frac{L_1^2}{2I_1} + \frac{L_2^2}{2I_2}. \qquad (3.10.20)
$$

Then the partition function becomes

$$
\begin{aligned}
Z_{\text{class}} &= \int_{-\infty}^{\infty} dL_1 dL_2 \exp\left(-\beta E_{\text{rot}}\right) \\
&= \int_{-\infty}^{\infty} dL_1 \exp\left(-\frac{\beta L_1^2}{2I_1}\right) \times \int_{-\infty}^{\infty} dL_2 \exp\left(-\frac{\beta L_2^2}{2I_2}\right) \\
&= \sqrt{\frac{2I_1}{\beta} \times \frac{2I_2}{\beta}} \left(\int_{-\infty}^{\infty} dx\, e^{-x^2}\right)^2 \propto \beta^{-1}.
\end{aligned} \qquad (3.10.21)
$$

This leads to

$$
U = Nk_BT \qquad (3.10.22)
$$

in accordance with Eq. (3.10.15) with $\xi = 1$.

3.11 Statistical physics of vibrations in solids

In Section 3.9, we have studied the statistical mechanics of harmonic oscillators. For instance, a diatomic molecule behaves as an oscillator, the chemical bond between the atoms periodically changing its length with time in the classical description. For molecules consisting of $N \geq 2$ atoms, vibrations become more complicated and, as a rule, involve all N atoms. Still, within the harmonic approximation (the potential energy expanded up to the second-order terms in deviations from the equilibrium positions), the classical equations of motion are linear and can be dealt with matrix algebra. Diagonalizing the dynamical matrix of the system, one can find $3N - 6$ linear combinations of atomic coordinates that dynamically behave as *independent* harmonic oscillators. Such *collective* vibrational modes of the system are called *normal modes*. For instance, a hypothetical triatomic molecule with atoms that are allowed to move along a straight line has two normal modes. If the masses of the first and third atoms are the same, one of the normal modes corresponds to antiphase oscillations of the first and third atoms with the second (central) atom being at rest. Another normal mode corresponds to the in-phase oscillations of the first and third atoms, with the central atom oscillating in the antiphase to them. For more complicated molecules, normal modes can be found numerically.

Normal modes can be considered quantum mechanically as independent quantum oscillators, as in Section 3.9. Then the vibrational partition function of the molecule is a product of partition functions of individual normal modes, and the internal energy is a sum of internal energies of these modes. The latter is the consequence of the independence of the normal modes.

The solid having a well-defined crystal structure is translationally invariant, which simplifies finding normal modes, in spite of a large number of atoms N. In the simplest case of one atom in the unit cell, normal modes can be found immediately by making a Fourier transformation of atomic deviations. The resulting normal modes are sound waves (or *phonons* in the quantum language) with different wave vectors **k** and polarizations (in the simplest case, one longitudinal and two equivalent transverse modes). In the sequel, to illustrate basic principles, we will consider only one type of sound waves (say, the longitudinal wave), but we will count it three times.

The results can then be generalized for different kinds of waves. For wave vectors small enough, there is a linear relation between the wave vector and the frequency of the oscillations

$$\omega_k = vk. \tag{3.11.1}$$

In this formula, ω_k depends only on the magnitude $k = 2\pi/\lambda$ and not on the direction of $\mathbf{k}$ and v is the speed of sound. The acceptable values of $\mathbf{k}$ should satisfy the boundary conditions at the boundaries of the solid body. Obviously, the thermodynamics of a large body is insensitive to the shape of the body and to the exact form of these conditions. (Note that the body's surface can be rough, and then it is difficult to formulate a model of what exactly happens at the surface!) The simplest possibility is to consider a body of the box shape with dimensions L_x, L_y, and L_z and use the clamped boundary conditions, so that the atoms at the boundary cannot move. These boundary conditions are the same as in the quantum problem of a particle in a potential box. One obtains the atomic displacements in the normal modes of the type $u(x) = u_0 \sin(k_x x)$, etc., and the discrete values of the wave vector components k_α with $\alpha = x, y, z$

$$k_\alpha = \pi \frac{\nu_\alpha}{L_\alpha}, \quad \nu = 1, 2, 3, \ldots \tag{3.11.2}$$

If the atoms are arranged in a simple cubic lattice with the lattice spacing a, then $L_\alpha = aN_\alpha$, where N_α is the number of atoms (or unit cells) in each direction α. Then

$$k_\alpha = \frac{\pi}{a} \frac{\nu_\alpha}{N_\alpha}. \tag{3.11.3}$$

Similar to the problem of the particle in a box, one can introduce the frequency density of normal modes $\rho(\omega)$ via the number of normal modes in the frequency interval $d\omega$

$$dn_\omega = \rho(\omega) \, d\omega. \tag{3.11.4}$$

Using Eq. (3.7.18) multiplied by the number of the phonon modes 3 and changing from k to ω with the help of Eq. (3.11.1), one obtains

$$\rho(\omega) = \frac{3V}{2\pi^2 v^3} \omega^2. \tag{3.11.5}$$

The total number of normal modes should be $3N$:

$$3N = 3\sum_{\mathbf{k}} 1 = 3 \sum_{\nu_x,\nu_y,\nu_z} 1. \qquad (3.11.6)$$

This equation splits into three separate equations:

$$N_\alpha = \sum_{\nu_\alpha=1}^{\nu_{\alpha,\max}} 1 = \nu_{\alpha,\max}. \qquad (3.11.7)$$

This defines the maximal values of ν_α and, according to Eq. (3.11.3), of k_α that happens to be $k_{\alpha,\max} = \pi/a$ for all α.

As the solid body is macroscopic and there are very many values of the allowed wave vectors, one can replace summation by integration and use the density of states:

$$\sum_{\mathbf{k}} \ldots \Rightarrow \int_0^{\omega_{\max}} d\omega \rho(\omega) \ldots \qquad (3.11.8)$$

At large k and thus ω, the linear dependence of Eq. (3.11.1) is no longer valid. Moreover, for models with elastic bonds on the lattice at large k, there are no purely longitudinal and transverse modes, and the mode frequency depends on the direction of the wave vector. Thus, Eq. (3.11.5) becomes invalid as well. Still, one can make a crude approximation, assuming that Eqs. (3.11.1) and (3.11.5) are valid until some maximal frequency of the sound waves in the body. The model based on this assumption is called the Debye model. The maximal frequency (or the Debye frequency ω_D) is defined by Eq. (3.11.6) that takes the form

$$3N = \int_0^{\omega_D} d\omega \rho(\omega). \qquad (3.11.9)$$

With the use of Eq. (3.11.5), one obtains

$$3N = \int_0^{\omega_D} d\omega \frac{3V}{2\pi^2 v^3} \omega^2 = \frac{V\omega_D^3}{2\pi^2 v^3}, \qquad (3.11.10)$$

where

$$\omega_D = \left(6\pi^2\right)^{1/3} \frac{v}{a}, \qquad (3.11.11)$$

where $V/N = v_0 = a^3$ is the unit-cell volume. It is convenient to rewrite Eq. (3.11.5) in terms of ω_D as

$$\rho\left(\omega\right) = 9N\frac{\omega^2}{\omega_D^3}. \tag{3.11.12}$$

The consideration up to this point is completely classical. Now, we will consider each **k**-mode as an independent harmonic oscillator. The quantum energy levels of the latter are given by Eq. (3.9.3). In our case for a **k**-oscillator, it becomes

$$\varepsilon_{\mathbf{k},\nu_{\mathbf{k}}} = \hbar\omega_k\left(\nu_{\mathbf{k}} + \frac{1}{2}\right), \quad \nu_{\mathbf{k}} = 0, 1, 2, \ldots \tag{3.11.13}$$

One quantum of the energy $\hbar\omega_k$ in the **k**-mode is called a phonon. The quantum number $\nu_{\mathbf{k}}$ is the number of phonons in the **k**-mode. Note the relation

$$\varepsilon = \hbar\omega \tag{3.11.14}$$

between the energy of a quantum of any wave (phonons, photons) and its frequency.

All the modes contribute additively to the internal energy, thus

$$U = 3\sum_{\mathbf{k}} \langle\varepsilon_{\mathbf{k},\nu_{\mathbf{k}}}\rangle = 3\sum_{\mathbf{k}} \hbar\omega_k\left(\langle\nu_{\mathbf{k}}\rangle + \frac{1}{2}\right) = 3\sum_{\mathbf{k}} \hbar\omega_k\left(n_{\mathbf{k}} + \frac{1}{2}\right), \tag{3.11.15}$$

where, similar to Eq. (3.9.19),

$$n_{\mathbf{k}} = \frac{1}{\exp\left(\beta\hbar\omega_k\right) - 1}. \tag{3.11.16}$$

Replacing summation by integration and rearranging Eq. (3.11.15), within the Debye model, one obtains

$$U = U_0 + \int_0^{\omega_D} d\omega \rho\left(\omega\right) \frac{\hbar\omega}{\exp\left(\beta\hbar\omega\right) - 1}, \tag{3.11.17}$$

where U_0 is the zero-temperature value of U due to the zero-point motion. The integral term in Eq. (3.11.17) can be calculated analytically at low and high temperatures.

For $k_B T \ll \hbar \omega_D$, the integral converges at $\omega \ll \omega_D$, so that one can extend the upper limit of integration to ∞. One obtains

$$
\begin{aligned}
U &= U_0 + \int_0^\infty d\omega \rho(\omega) \frac{\hbar \omega}{\exp(\beta \hbar \omega) - 1} \\
&= U_0 + \frac{9N}{\omega_D^3} \int_0^\infty d\omega\, \omega^2 \frac{\hbar \omega}{\exp(\beta \hbar \omega) - 1} \\
&= U_0 + \frac{9N}{(\hbar \omega_D)^3 \beta^4} \int_0^\infty dx \frac{x^3}{e^x - 1} = U_0 + \frac{9N}{(\hbar \omega_D)^3 \beta^4} \frac{\pi^4}{15} \\
&= U_0 + \frac{3\pi^4}{5} N k_B T \left(\frac{k_B T}{\hbar \omega_D} \right)^3 .
\end{aligned}
\tag{3.11.18}
$$

This result can be written in the form

$$
U = U_0 + \frac{3\pi^4}{5} N k_B T \left(\frac{T}{\theta_D} \right)^3 ,
\tag{3.11.19}
$$

where

$$
\theta_D = \hbar \omega_D / k_B
\tag{3.11.20}
$$

is the Debye temperature. In fact, here, the Debye temperature is only a convenient parameter to represent the result that is insensitive to the assumption made in the Debye theory. At low temperatures, the internal energy does not depend on the underlying lattice structure of the solid and can be rewritten in the form of the macroscopic continuous theory:

$$
U = U_0 + \frac{\pi^2}{10} \frac{(k_B T)^4}{\hbar^3 v^3} V.
\tag{3.11.21}
$$

The heat capacity is given by

$$
C_V = \left(\frac{\partial U}{\partial T} \right)_V = \frac{12\pi^4}{5} N k_B \left(\frac{T}{\theta_D} \right)^3 , \quad T \ll \theta_D.
\tag{3.11.22}
$$

For $k_B T \gg \hbar \omega_D$, the exponential in Eq. (3.11.17) can be expanded for $\beta \hbar \omega \ll 1$. It is, however, convenient to step back and to do the same in Eq. (3.11.15). With the help of Eq. (3.11.6), one obtains

$$U = U_0 + 3 \sum_{\mathbf{k}} \frac{\hbar \omega_k}{\beta \hbar \omega_k} = U_0 + 3k_B T \sum_{\mathbf{k}} 1 = U_0 + 3Nk_B T. \qquad (3.11.23)$$

Note that the phonon spectrum ω_k, which is due to the interactions in the system, disappeared from the formula. For the heat capacity, one then obtains

$$C_V = 3Nk_B, \qquad (3.11.24)$$

k_B per each of approximately $3N$ vibrational modes. One can see that Eqs. (3.11.23) and (3.11.24) are accurate because they do not use the Debye model. The assumption of the Debye model is essential only at intermediate temperatures, where Eq. (3.11.17) is approximate.

The temperature dependence of the phonon heat capacity, following from Eq. (3.11.17), is shown in Fig. 3.4 together with its low temperature asymptote, Eq. (3.11.22), and the high-temperature asymptote, Eq. (3.11.24). Due to the large numerical coefficient in Eq. (3.11.22), the law $C_V \propto T^3$ in fact holds only for temperatures much smaller than the Debye temperature.

The theory of quantized vibrations in the solid, developed by Debye, improves on the more primitive Einstein model that assumes independent oscillators.

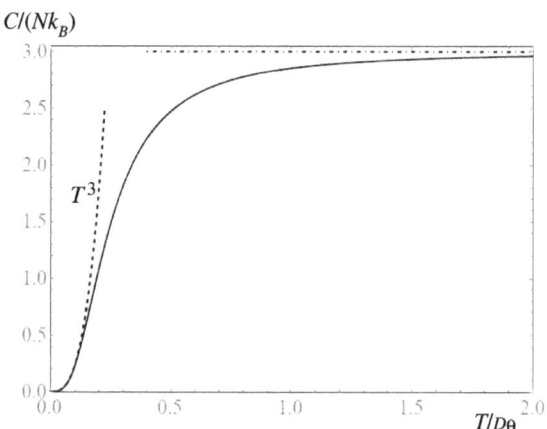

Figure 3.4. Temperature dependence of the phonon heat capacity. The low temperature result $C \propto T^3$ (dashed line) holds only for temperatures much smaller than the Debye temperature θ_D.

3.12 Problems

3.12.1 Microstates and macrostates

Three distinguishable particles can occupy four states. Find all macrostates and the numbers of microstates realizing each macrostate. Put the results into the table. Which macrostate has the highest statistical weight? What is the total number of microstates that can be found immediately? Is the sum of the microstates realizing each macrostate equal to this expected total number?

Macrostates			Number of microstates
3	0	0	1
...	...	...	...

Solution: The table is as follows:

Macrostates			Number of microstates
3	0	0	1
0	3	0	1
0	0	3	1
2	1	0	3
2	0	1	3
1	2	0	3
1	0	2	3
0	2	1	3
0	1	2	3
1	1	1	6

Note: The total number of microstates is $3^3 = 27$.

3.12.2 Method of Lagrange multipliers

Find the area of the largest rectangle that can be inscribed in the ellipse:

$$\frac{x^2}{a^2} + \frac{y^2}{b^2} = 1.$$

Use the method of Lagrange multipliers: To minimize a function $F(x, y)$ with a constraint $\phi(x, y) = 0$, minimize

$$\Phi(x, y) \equiv F(x, y) - \lambda\phi(x, y)$$

with respect to x, y, and λ.

Solution: Using the method of Lagrange multipliers, we minimize

$$\Phi(x, y) = xy - \lambda\left(\frac{x^2}{a^2} + \frac{y^2}{b^2} - 1\right).$$

The equations for the extrema are

$$0 = \frac{\partial\Phi}{\partial x} = y - 2\lambda x/a^2,$$

$$0 = \frac{\partial\Phi}{\partial y} = x - 2\lambda y/b^2,$$

$$0 = \frac{\partial\Phi}{\partial \lambda} = \frac{x^2}{a^2} + \frac{y^2}{b^2} - 1.$$

Expressing

$$y = 2\lambda x/a^2$$

from the first equation and inserting the result into the second equation, one obtains

$$0 = x - 4\lambda^2 x/(ab)^2,$$

i.e.,

$$\lambda = ab/2.$$

Then one obtains

$$y = x\frac{b}{a}.$$

Inserting this into the third extremum equation, one obtains

$$0 = \frac{x^2}{a^2} + \frac{x^2}{a^2} - 1$$

and

$$x = a/\sqrt{2}, \quad y = b/\sqrt{2}.$$

The maximal area is now

$$A_{\max} = xy = \frac{ab}{2}.$$

A posteriori, one can see that with the smart choice of variables

$$\tilde{x} \equiv x/a, \quad \tilde{y} \equiv y/b,$$

one could achieve a more elegant and probably completely symmetric solution of the equations.

3.12.3 Thermodynamics of particles in a rigid box in the quantum limit

Calculate the partition function, the internal energy, and the heat capacity of an ensemble of noninteracting particles in a cubic box $L_x = L_y = L_z = L$ at low temperatures (quantum limit), assuming Boltzmann statistics. Define the crossover temperature between the quantum and classical limits.

Solution: The quantum energy levels of the particle in a cubic box are given by

$$\varepsilon_{\nu_x,\nu_y,\nu_z} = \frac{\pi^2 \hbar^2}{2mL^2} \left(\nu_x^2 + \nu_y^2 + \nu_z^2 \right), \quad \nu_\alpha = 1, 2, 3, \ldots \qquad (3.12.1)$$

The energy of the ground state $(\nu_x, \nu_y, \nu_z) = (1, 1, 1)$ is

$$\varepsilon_{111} = \frac{3\pi^2 \hbar^2}{2mL^2}. \qquad (3.12.2)$$

The energy of the triple-degenerate first excited states $(2, 1, 1)$, $(1, 2, 1)$, $(1, 1, 2)$ is

$$\varepsilon_{211} = \varepsilon_{121} = \varepsilon_{112} = \frac{6\pi^2 \hbar^2}{2mL^2}.$$

The partition function is given by

$$\mathcal{Z} = \exp\left(-\beta \frac{3\pi^2 \hbar^2}{2mL^2} \right) + 3\exp\left(-\beta \frac{6\pi^2 \hbar^2}{2mL^2} \right) + \cdots$$

or

$$\mathcal{Z} = \exp\left(-\beta\frac{3\pi^2\hbar^2}{2mL^2}\right)\left[1 + 3\exp\left(-\beta\frac{3\pi^2\hbar^2}{2mL^2}\right) + \cdots\right].$$

The terms denoted by $\cdots$ are contributions of the higher levels. In the low temperature range,

$$k_BT \ll k_BT_0 \equiv U_0 \equiv \frac{3\pi^2\hbar^2}{2mL^2},$$

the ground state makes the dominant contribution to the partition function, while the contributions of the excited states are exponentially small. To obtain the temperature-dependent part of the thermodynamic quantities, it is sufficient to keep the contribution of the first excited states because those from higher excited states are much smaller. From the expression above, one obtains

$$\ln\mathcal{Z} = -\beta U_0 + \ln[1 + 3\exp(-\beta U_0) + \cdots] \cong -\beta U_0 + 3\exp(-\beta U_0).$$

The internal energy is given by

$$U = -N\frac{\partial\ln\mathcal{Z}}{\partial\beta} = NU_0 + 3NU_0\exp\left(-\beta U_0\right)$$

$$= NU_0[1 + 3\exp(-\beta U_0)].$$

The first term is just the ground-state energy. The second term is the exponentially small thermal energy. The heat capacity is entirely due to the latter:

$$C = \frac{dU}{dT} = 3NU_0\left[\frac{\partial}{\partial\beta}\exp\left(-\beta U_0\right)\right]\frac{\partial\beta}{\partial T}$$

$$= 3Nk_B\left(\frac{U_0}{k_BT}\right)^2\exp\left(-\frac{U_0}{k_BT}\right)$$

or

$$C = 3Nk_B\left(\frac{T_0}{T}\right)^2\exp\left(-\frac{T_0}{T}\right) \ll Nk_B.$$

The temperature T_0 defined above is the crossover temperature between the quantum and classical regimes. For $T \gtrsim T_0$, many excited levels make a contribution to the partition function, and one has to perform a numerical summation to compute $\mathcal{Z}$. For $T \gg T_0$, the highly excited states are dominating, and one can replace summation by integration and calculate the partition function analytically.

3.12.4 Density of states of particles in rigid 1D and 2D boxes

The density of states of quantum particles in a rigid 3D box has been calculated in the lectures. Generalize these results for one and two dimensions.

Solution: In one dimension, the energy levels are given by

$$\varepsilon_n = \frac{\hbar^2 k_\nu^2}{2m}, \quad k_\nu = \frac{\pi}{L}\nu, \quad \nu = 1, 2, 3, \ldots \tag{3.12.3}$$

To calculate the density of states defined by

$$dn_\varepsilon = \rho(\varepsilon)d\varepsilon, \tag{3.12.4}$$

start with

$$dn_\nu = d\nu \tag{3.12.5}$$

as the number of states (energy levels) in the interval $d\nu$ of the quantum number ν. Using the expression for the quantized wave vector in Eq. (3.12.3), one can rewrite Eq. (3.12.5) in terms of k as

$$dn_k = \frac{L}{\pi}dk.$$

Using now the relation between the wave vector and the energy,

$$k = \sqrt{\frac{2m\varepsilon}{\hbar^2}}, \tag{3.12.6}$$

one obtains

$$dn_\varepsilon = \frac{L}{\pi}\frac{dk}{d\varepsilon}d\varepsilon = \frac{L}{2\pi}\left(\frac{2m}{\hbar^2}\right)^{1/2}\frac{d\varepsilon}{\sqrt{\varepsilon}}.$$

Thus, the density of states is given by

$$\rho(\varepsilon) = \frac{L}{2\pi}\left(\frac{2m}{\hbar^2}\right)^{1/2}\frac{1}{\sqrt{\varepsilon}}$$

In two dimensions, instead of Eq. (3.12.5), we use

$$dn_{\nu_x \nu_y} = d\nu_x d\nu_y, \tag{3.12.7}$$

which in terms of k becomes

$$dn_{k_x k_y} = \frac{L_x L_y}{\pi^2} dk_x dk_y. \tag{3.12.8}$$

Now, we go over to the number of states within the circular shell $k, k + dk$, taking into account that both k_x and k_y are positive, and thus, there is only one quarter of the circular shell. One obtains

$$dn_k = \frac{L_x L_y}{\pi^2} \frac{2\pi}{4} k dk.$$

Further, with the help of Eq. (3.12.6), follows

$$dn_\varepsilon = \frac{S}{2\pi} \sqrt{\frac{2m\varepsilon}{\hbar^2}} \left(\frac{2m}{\hbar^2}\right)^{1/2} \frac{d\varepsilon}{\sqrt{\varepsilon}} = \frac{S}{2\pi} \frac{2m}{\hbar^2} d\varepsilon,$$

and thus,

$$\rho(\varepsilon) = \frac{S}{2\pi} \frac{2m}{\hbar^2},$$

where $S = L_x L_y$ is the area of the rigid box. Note that in 2D particle's density of states is a constant.

3.12.5   Density of states of 1D and 2D phonons

The density of states of 3D phonons has been calculated in the lectures. Generalize these results for one and two dimensions.

Solution: In one dimension, we use the density of states with respect to the wave vector k, which is the same for particles and lattice vibrations:

$$dn_k = \frac{L}{\pi} dk.$$

Here, we change from k to ω using the phonon dispersion law $\omega = vk$ to obtain the density of states

$$\rho(\omega) = \frac{L}{\pi v},$$

which is a constant.

In two dimensions one can start, again, with the DOS in terms of k:

$$dn_k = \frac{S}{2\pi} k\, dk,$$

as for the particles above. Here, $S = L_x L_y$ is the area of the system. Changing to ω, one obtains the phonon DOS:

$$\rho(\omega) = \frac{S\omega}{2\pi v^2}.$$

3.12.6 Statistical thermodynamics of free classical particles

Partition function of classical particles in 3D is defined as

$$\mathcal{Z}_{\text{class}} = \int d^3p \int d^3r \exp\left[-\beta E(\mathbf{p}, \mathbf{r})\right], \qquad (3.12.9)$$

where $E(\mathbf{p}, \mathbf{r})$ is the particle's energy. Note that this expression has the unit of (momentum × distance)^3, unlike the quantum partition function, which is dimensionless. Define the density of states of a free classical particle in a box of volume V. By comparing it with the density of states for a quantum particle in a rigid box, find the missing factor in Eq. (3.12.9) that would make the classical partition function match the quantum one. This will define a quantum-mechanical "cell" in the phase space of a classical particle. Show that this quantum-mechanical aspect does not contribute to the internal energy and heat capacity of the classical particles.

Solution: The energy of the particle consists of the kinetic and potential energies,

$$E(\mathbf{p}, \mathbf{r}) = \frac{\mathbf{p}^2}{2m} + U(\mathbf{r}), \qquad (3.12.10)$$

so that the classical partition function factorizes:

$$\mathcal{Z}_{\text{class}} = \int d^3p\, e^{-\beta p^2/(2m)} \int d^3r\, e^{-\beta U(\mathbf{r})}. \qquad (3.12.11)$$

For free particles, there is no potential energy, and $\mathcal{Z}_{\text{class}}$ for particles in a rigid box of volume V becomes

$$\mathcal{Z}_{\text{class}} = V \int d^3p\, e^{-\beta p^2/(2m)}. \qquad (3.12.12)$$

In the spherical coordinate system, this becomes

$$\mathcal{Z}_{\text{class}} = 4\pi V \int p^2 dp\, e^{-\beta p^2/(2m)}. \qquad (3.12.13)$$

Choosing the kinetic energy $\varepsilon = p^2/(2m)$ as the integration variable and using

$$p^2 = 2m\varepsilon, \quad dp = \frac{dp}{d\varepsilon} d\varepsilon = \frac{1}{2}\sqrt{\frac{2m}{\varepsilon}} d\varepsilon, \qquad (3.12.14)$$

one can rewrite this in the form

$$\mathcal{Z}_{\text{class}} = \int_0^\infty d\varepsilon\, \rho_{\text{class}}(\varepsilon) e^{-\beta\varepsilon}, \quad \rho_{\text{class}}(\varepsilon) = 2\pi V (2m)^{3/2} \sqrt{\varepsilon}.$$

Quantum-mechanical partition function for this problem has the same form with

$$\rho(\varepsilon) = \frac{V}{(2\pi)^2} \left(\frac{2m}{\hbar^2}\right)^{3/2} \sqrt{\varepsilon}. \qquad (3.12.15)$$

The two densities of states are related as

$$\rho(\varepsilon) = \frac{\rho_{\text{class}}(\varepsilon)}{(2\pi\hbar)^3},$$

which defines the missing factor in the above definition of the classical partition function. Correcting $\mathcal{Z}_{\text{class}}$ as

$$\mathcal{Z}_{\text{class}} = \int \frac{d^3p\, d^3r}{(2\pi\hbar)^3} \exp\left[-\beta E(\mathbf{p}, \mathbf{r})\right], \qquad (3.12.16)$$

one obtains the dimensionless quantity that coincides with the classical limit of the quantum partition function Z. This formula can be used in the presence of the potential energy as well.

The interpretation of the above is the following. The (x, p_x) projection of the phase space of the particle is discretized into cells, $\Delta x \Delta p_x = 2\pi\hbar = h$, and similar for other directions y and z. The cells have a quantum origin and are related to Heisenberg's uncertainty principle, stating that the product of uncertainties of measuring x and p_x of a quantum particle is of order h. The number of quantum cells in a limited region of x and p_x is limited, and it defines the number of different states in this region. It is impossible to have more different states because there is no way to distinguish states that are too close both in x and in p_x by measurement.

Similar, quantum cells can be introduced in many-particle problems and in problems with rotational degrees of freedom. Quantum cell is an external element in classical statistical physics. Statistical averages of most physical quantities (except for the entropy and related functions) are insensitive to the quantization of the phase space of the system because the correction factor introduced above cancels.

3.12.7 Classical particles with gravity

Using the distribution function

$$f(\mathbf{p}, \mathbf{r}) = \frac{1}{\mathcal{Z}_{\text{class}}} \exp\left[-\beta E(\mathbf{p}, \mathbf{r})\right]$$

for classical particles with gravity, find the dependence of particles' concentration n and pressure P as a function of the height. Set the minimal height (the ground level) to zero. Calculate the heat capacity of this system and compare it with the one for free particles.

Solution: The energy of the particle has the form

$$E = \frac{p^2}{2m} + mgz.$$

The classical partition function of the particle factorizes:

$$\mathcal{Z}_{\text{class}} = \int d^3p \int d^3r \exp\left[-\beta E(\mathbf{p}, \mathbf{r})\right] = \mathcal{Z}_{\text{kinetic}} \mathcal{Z}_{\text{potential}},$$

where

$$\mathcal{Z}_{\text{kinetic}} = \int d^3p\, e^{-\beta p^2/(2m)}, \quad \mathcal{Z}_{\text{potential}} = \int d^3r\, e^{-\beta mgz}.$$

Here, we do not use the phase-space quantization factor from the preceding problem because it will cancel out in the final results.

The kinetic part of Z can be calculated as

$$\mathcal{Z}_{\text{kinetic}} = \left[\int_{-\infty}^{\infty} dp_x e^{-\beta p_x^2 / (2m)} \right]^3 = (2\pi m k_B T)^{3/2} .$$

Assuming that the particles are contained in a vertical cylinder of the cross-section S (that is nonessential), for the potential partition function, one obtains

$$\mathcal{Z}_{\text{potential}} = S \int_0^{\infty} dz e^{-\beta m g z} = \frac{S k_B T}{m g} .$$

Thus,

$$\mathcal{Z}_{\text{class}} = \frac{S}{g} (2\pi)^{3/2} m^{1/2} (k_B T)^{5/2} .$$

Suppose there are N particles in the system. Then the number of particles in the element of the phase space is

$$dN = N f dp_x dp_y dp_z dx dy dz,$$

where f is the distribution function introduced above. The concentration of particles is defined as

$$n = \int d^3 p \frac{dN}{dx dy dz} = N \int d^3 p f.$$

Since f and $\mathcal{Z}_{\text{class}}$ factorize, the integrals over momentum cancel out, and one obtains

$$n = \frac{N}{\mathcal{Z}_{\text{potential}}} e^{-\beta m g z} .$$

One can check that integrating this over the volume yields the identity $N = N$. Using the result for $\mathcal{Z}_{\text{potential}}$, one gets the final result

$$n = \frac{m g N}{S k_B T} e^{-\beta m g z},$$

which exponentially decreases with the height. For the pressure, one does not have anything better than the ideal-gas formula

$$P = n k_B T = \frac{m g N}{S} e^{-\beta m g z} .$$

The average internal energy is given by

$$U = -N \frac{\partial \ln \mathcal{Z}_{\text{class}}}{\partial \beta} = -N \frac{\partial \ln \beta^{-5/2}}{\partial \beta} = \frac{5}{2} \frac{N}{\beta} = \frac{5}{2} N k_B T.$$

The heat capacity is

$$C = \frac{\partial U}{\partial T} = \frac{5}{2} N k_B.$$

This result might be unexpected. There is the heat capacity $(3/2)Nk_B$ from three translational degrees of freedom. Additionally, there is a potential energy for the motion in the vertical direction. Its contribution is Nk_B instead of the expected $(1/2)Nk_B$, as was the case for a vibrational degree of freedom. The reason for a different result is that the theorem of the equidistribution of the energy over degrees of freedom is valid in the cases when the energy is a quadratic function in momenta and deviations from the equilibrium positions (see the next problem). In our case, the potential energy is a linear rather than a quadratic function of z.

3.12.8 Classical 3D harmonic oscillators

Consider classical particles with the potential energy

$$V(\mathbf{r}) = \frac{kr^2}{2}$$

in three dimensions. Calculate the partition function, internal energy, and heat capacity.

Solution: We start, as usual, with calculating the classical partition function

$$\mathcal{Z}_{\text{class}} = \int d^3 p \, e^{-\beta p^2/(2m)} \int d^3 r \, e^{-\beta kr^2/2} = \mathcal{Z}_{\text{kinetic}} \mathcal{Z}_{\text{potential}}.$$

$$(3.12.17)$$

One obtains

$$\mathcal{Z}_{\text{kinetic}} = \left[\int_{-\infty}^{\infty} dp_x e^{-\beta p_x^2/(2m)} \right]^3 = (2\pi m k_B T)^{3/2}.$$

Similar,

$$\mathcal{Z}_{\text{potential}} = \left[\int_{-\infty}^{\infty} dx e^{-\beta k x^2/2} \right]^3 = (2\pi k_B T / k)^{3/2}.$$

Thus,

$$\mathcal{Z}_{\text{class}} \propto T^3 \propto \beta^{-3}.$$

The internal energy and heat capacity are given by

$$U = -N \frac{\partial \ln \mathcal{Z}_{\text{class}}}{\partial \beta} = \frac{3N}{\beta} = 3N k_B T$$

and

$$C = \frac{\partial U}{\partial T} = 3N k_B.$$

The factor 3 here is due to the three translational degrees of freedom of our system. Per degree of freedom, there is $1/2 N k_B$ due to the kinetic energy and the same amount due to the potential energy. This problem illustrates the equidistribution of energy over degrees of freedom in classical statistical physics.

3.12.9 1D and 2D phonons

Calculate the internal energy and heat capacity of the system of harmonic phonons in one and two dimensions at low temperatures.

Solution: Instead of calculating the partition function, it is more convenient to use the direct formula for the internal energy:

$$U = U_0 + \int_0^{\infty} d\omega \rho(\omega) \frac{\hbar \omega}{\exp(\beta \hbar \omega) - 1}, \qquad (3.12.18)$$

where the upper limit of the integration has been set to infinity at low temperatures. In two dimensions, the phonon density of states is

$$\rho(\omega) = \frac{S\omega}{2\pi v^2},$$

where S is the system's area. Discarding the zero-point energy U_0, one obtains

$$U = \frac{S\hbar}{2\pi v^2} \int_0^\infty d\omega \frac{\omega^2}{\exp(\beta \hbar \omega) - 1}$$

$$= \frac{S\hbar}{2\pi v^2} \left(\frac{k_B T}{\hbar}\right)^3 \int_0^\infty dx \frac{x^2}{e^x - 1} = \frac{\zeta(3) S (k_B T)^3}{\pi v^2 \hbar^2}.$$

The heat capacity becomes

$$C = \frac{\partial U}{\partial T} = 3k_B \frac{\zeta(3) S (k_B T)^2}{\pi v^2 \hbar^2}.$$

3.12.10 Quantum correction to the heat capacity of the harmonic oscillator at high temperatures

Calculate the first quantum correction to the heat capacity of the harmonic oscillator at high temperatures.

Solution: The formula for the heat capacity of the ensemble of harmonic oscillators reads

$$C = N k_B \left(\frac{\beta \hbar \omega_0}{\sinh[\beta \hbar \omega_0]}\right)^2. \qquad (3.12.19)$$

At high temperatures (small β), one can expand this expression using the formula

$$\sinh x \cong x + \frac{x^3}{6}, \quad x \ll 1.$$

One writes

$$\left(\frac{x}{\sinh x}\right)^2 \cong \left(\frac{1}{1 + x^2/6}\right)^2 \cong \left(1 - x^2/6\right)^2 \cong 1 - x^2/3.$$

Thus, in the high-temperature range,

$$C \cong Nk_B \left[1 - \frac{(\beta\hbar\omega_0)^2}{3} \right] = Nk_B \left[1 - \frac{1}{3} \left(\frac{\hbar\omega_0}{k_B T} \right)^2 \right].$$

The quantum correction contains $\hbar$ and vanishes in the classical limit $\hbar \to 0$.

3.13 Spins in the magnetic field

Magnetism of solids is in most cases due to the intrinsic angular momentum of the electron that is called *spin*. In quantum mechanics, spin is an operator represented by a matrix. The value of the electronic angular momentum is $s = 1/2$, so that it is described by a 2×2 matrix and the quantum-mechanical eigenvalue of the square of the electron's angular momentum (in the units of $\hbar$) is

$$\hat{s}^2 = s\,(s+1) = 3/4, \qquad (3.13.1)$$

c.f. Eq. (3.10.4). Eigenvalues of $\hat{s}_z$, projection of a free spin on any direction z, are given by

$$\hat{s}_z = m, \quad m = \pm 1/2, \qquad (3.13.2)$$

c.f. Eq. (3.10.5). The spin of the electron can be interpreted as the circular motion of the charge inside this elementary particle. This results in the magnetic moment of the electron

$$\hat{\boldsymbol{\mu}} = g\mu_B \hat{\mathbf{s}}, \qquad (3.13.3)$$

where $g = 2$ is the so-called g-factor and μ_B is Bohr's magneton, $\mu_B = e\hbar/(2m_e) = 0.927 \times 10^{-23}$ J/T. Note that the model of circularly moving charges inside the electron leads to the same result with $g = 1$; the true value $g = 2$ follows from the relativistic quantum theory.

The energy (the Hamiltonian) of an electron spin in a magnetic field of induction $\mathbf{B}$ has the form

$$\hat{H} = -\hat{\boldsymbol{\mu}} \cdot B = -g\mu_B \hat{\mathbf{s}} \cdot \mathbf{B}, \qquad (3.13.4)$$

the so-called Zeeman Hamiltonian. One can see that the minimum of the energy corresponds to the spin pointing in the direction of $\mathbf{B}$.

Choosing the z-axis in the direction of $\mathbf{B}$, one obtains

$$\hat{H} = -g\mu_B \hat{s}_z B. \tag{3.13.5}$$

Thus, the energy eigenvalues of the electronic spin in a magnetic field with the help of Eq. (3.13.2) become

$$\varepsilon_m = -g\mu_B m B, \tag{3.13.6}$$

where, for the electron, $m = \pm 1/2$.

In an atom, spins of all electrons usually combine into a collective spin S that has half-integer values and is described by a $(2S + 1) \times (2S + 1)$ matrix. Instead of Eq. (3.13.4), one has

$$\hat{H} = -g\mu_B \hat{\mathbf{S}} \cdot \mathbf{B}. \tag{3.13.7}$$

Eigenvalues of the projections of $\hat{\mathbf{S}}$ on the direction of $\mathbf{B}$ are given by

$$m = -S, -S + 1, \ldots, S - 1, S, \tag{3.13.8}$$

all together $2S + 1$ different values, c.f. Eq. (3.10.4). The $2S + 1$ different energy levels of the spin are given by Eq. (3.13.6).

The thermodynamics of the spin is determined by the partition function

$$Z_S = \sum_{m=-S}^{S} e^{-\beta \varepsilon_m} = \sum_{m=-S}^{S} e^{my}, \tag{3.13.9}$$

where ε_m is given by Eq. (3.13.6) and

$$y \equiv \beta g\mu_B B = \frac{g\mu_B B}{k_B T}. \tag{3.13.10}$$

Equation (3.13.9), by the change of the summation index to $k = m + S$, can be reduced to a finite geometrical progression

$$Z_S = e^{-Sy} \sum_{k=0}^{2S} (e^y)^k = e^{-Sy} \frac{e^{(2S+1)y} - 1}{e^y - 1} = \frac{\sinh\left[(S + 1/2)y\right]}{\sinh\left(y/2\right)}.$$

$$\tag{3.13.11}$$

For $S = 1/2$, with the help of $\sinh(2x) = 2\sinh(x)\cosh(x)$, one can transform Eq. (3.13.11) to

$$Z_{1/2} = 2\cosh(y/2). \tag{3.13.12}$$

The partition function being found, one can easily obtain thermodynamic quantities. For instance, the free energy F is given by Eq. (3.6.18):

$$F = -Nk_BT \ln Z_S = -Nk_BT \ln \frac{\sinh\left[(S + 1/2)y\right]}{\sinh(y/2)}. \tag{3.13.13}$$

The average spin polarization is defined by

$$\langle S_z \rangle = \langle m \rangle = \frac{1}{Z} \sum_{m=-S}^{S} m e^{-\beta\varepsilon m}. \tag{3.13.14}$$

With the help of Eq. (3.13.9), one obtains

$$\langle S_z \rangle = \frac{1}{Z}\frac{\partial Z}{\partial y} = \frac{\partial \ln Z}{\partial y}. \tag{3.13.15}$$

With the use of Eq. (3.13.11) and the derivatives

$$\sinh(x)' = \cosh(x), \quad \cosh(x)' = \sinh(x), \tag{3.13.16}$$

one obtains

$$\langle S_z \rangle = b_S(y), \tag{3.13.17}$$

where

$$b_S(y) = \left(S + \frac{1}{2}\right)\coth\left[\left(S + \frac{1}{2}\right)y\right] - \frac{1}{2}\coth\left[\frac{1}{2}y\right]. \tag{3.13.18}$$

Usually the function $b_S(y)$ is represented in the form $b_S(y) = SB_S(Sy)$, where $B_S(x)$ is the Brillouin function

$$B_S(x) = \left(1 + \frac{1}{2S}\right)\coth\left[\left(1 + \frac{1}{2S}\right)x\right] - \frac{1}{2S}\coth\left[\frac{1}{2S}x\right] \tag{3.13.19}$$

(see Fig. 3.5). One can see that $b_S(\pm\infty) = \pm S$ and $B_S(\pm\infty) = \pm 1$.

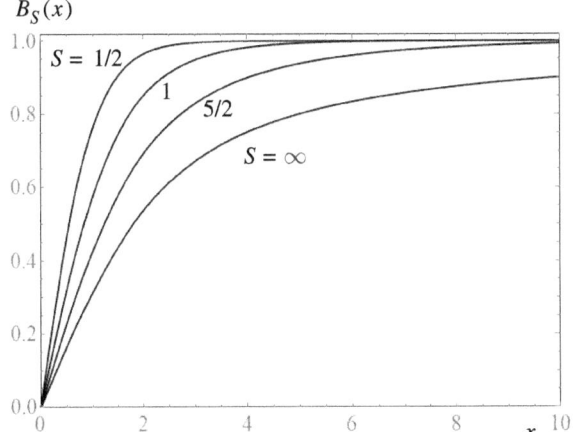

Figure 3.5. Brillouin function $B_S(x)$ for different S.

For $S = 1/2$, from Eq. (3.13.12), one obtains

$$\langle S_z \rangle = b_{1/2}(y) = \frac{1}{2} \tanh \frac{y}{2} \qquad (3.13.20)$$

and $B_{1/2}(x) = \tanh x$.

In the opposite limit, $S \to \infty$, the Brillouin function becomes the classical Langevin function:

$$B_{1/2}(x) \equiv L(x) = \coth x - \frac{1}{x}. \qquad (3.13.21)$$

Although atoms never have a really large spin, the classical approximation is very useful as it can strongly simplify calculations and provide qualitatively reasonable results in most cases.

At zero magnetic field, $y = 0$, the spin polarization should vanish. It is immediately seen in Eq. (3.13.20), but not in Eq. (3.13.17). To clarify the behavior of $b_S(y)$ at small y, one can use $\coth x \simeq 1/x + x/3$ that yields

$$b_S(y) \cong \left(S + \frac{1}{2} \right) \left[\frac{1}{\left(S + \frac{1}{2} \right) y} + \left(S + \frac{1}{2} \right) \frac{y}{3} \right] - \frac{1}{2} \left[\frac{1}{\frac{1}{2} y} + \frac{1}{2} \frac{y}{3} \right]$$

$$= \frac{y}{3} \left[\left(S + \frac{1}{2} \right)^2 - \left(\frac{1}{2} \right)^2 \right] = \frac{S(S+1)}{3} y. \qquad (3.13.22)$$

In physical units, this means

$$\langle S_z \rangle \cong \frac{S(S+1)}{3} \frac{g\mu_B B}{k_B T}, \quad g\mu_B B \ll k_B T. \tag{3.13.23}$$

The average magnetic moment of the spin can be obtained from Eq. (3.13.3):

$$\langle \mu_z \rangle = g\mu_B \langle S_z \rangle. \tag{3.13.24}$$

The magnetization of the sample M is defined as the magnetic moment per unit volume. If there is one magnetic atom per unit cell and all of them are uniformly magnetized, the magnetization reads

$$M = \frac{\langle \mu_z \rangle}{v_0} = \frac{g\mu_B}{v_0} \langle S_z \rangle, \tag{3.13.25}$$

where v_0 is the unit-cell volume.

The internal energy U of a system of N spins can be obtained from the general formula, Eq. (3.6.11). The calculation can, however, be avoided since from Eq. (3.13.6) simply follows

$$U = N \langle \varepsilon_m \rangle = -N g\mu_B B \langle m \rangle = -N g\mu_B B \langle S_z \rangle, \tag{3.13.26}$$

where $\langle S_z \rangle$ is given by Eq. (3.13.17). In particular, at low magnetic fields (or high temperatures), one has

$$U \cong -N \frac{S(S+1)}{3} \frac{(g\mu_B B)^2}{k_B T}. \tag{3.13.27}$$

The magnetic susceptibility per spin is defined by

$$\chi = \frac{\partial \langle \mu_z \rangle}{\partial B}. \tag{3.13.28}$$

From Eqs. (3.13.17) and (3.13.24), one obtains

$$\chi = \frac{(g\mu_B)^2}{k_B T} b_S'(y), \tag{3.13.29}$$

where

$$b_S'(y) \equiv \frac{db_S(y)}{dy} = -\left(\frac{S+1/2}{\sinh\left[(S+1/2)\,y\right]} \right)^2 + \left(\frac{1/2}{\sinh\left[y/2\right]} \right)^2. \tag{3.13.30}$$

For $S = 1/2$, from Eq. (3.13.20) follows

$$b'_{1/2}(y) = \frac{1}{4\cosh^2(y/2)}. \tag{3.13.31}$$

From Eq. (3.13.22), one obtain $b'_S(0) = S(S+1)/3$. Thus, in the high-temperature limit (or for $B = 0$),

$$\chi = \frac{S(S+1)}{3}\frac{(g\mu_B)^2}{k_BT}, \quad k_BT \gg g\mu_B B. \tag{3.13.32}$$

The susceptibility becomes small at high temperatures as the directions of the spins are strongly disordered by thermal agitation, and it is difficult to order them by applying a magnetic field. In the opposite limit, $y \gg 1$, the function $b'_S(y)$ and thus the susceptibility also become small. The physical reason for this is that at low temperatures, $k_BT \ll g\mu_B B$, the spins are already strongly aligned by the magnetic field, $\langle S_z \rangle \cong S$, so that $\langle S_z \rangle$ becomes hardly sensitive to small changes of B. As a function of temperature, χ has a maximum at intermediate temperatures.

The heat capacity C can be obtained from Eq. (3.13.26) as

$$C = \frac{\partial U}{\partial T} = -Ng\mu_B B\frac{\partial \langle S_z \rangle}{\partial T} = -Ng\mu_B Bb'_S(y)\frac{\partial y}{\partial T} = Nk_B y^2 b'_S(y). \tag{3.13.33}$$

As both the magnetic susceptibility and the heat capacity are expressed via the derivative of the Brillouin function, they are related. The relation has the form

$$\frac{C}{N\chi} = \frac{B^2}{T}. \tag{3.13.34}$$

One can see that C also has a maximum at intermediate temperatures, $y \sim 1$. At high temperatures, C decreases as $1/T^2$:

$$C \cong Nk_B\frac{S(S+1)}{3}\left(\frac{g\mu_B B}{k_BT}\right)^2, \tag{3.13.35}$$

and at low temperatures, C becomes exponentially small, except for the case $S = \infty$ (see Fig. 3.6). The finite value of the heat capacity in the limit $T \to 0$ contradicts the third law of thermodynamics. Thus, the classical spin model fails in this limit.

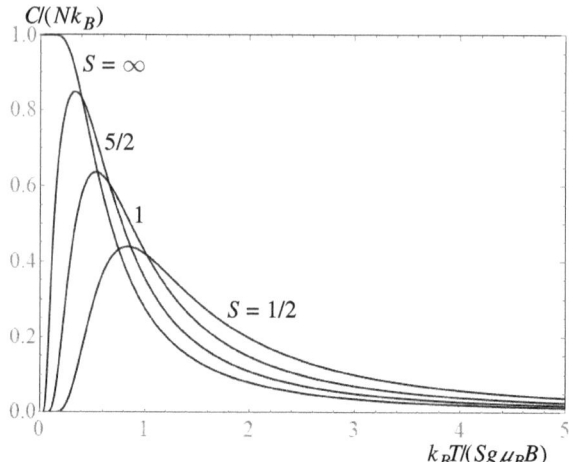

Figure 3.6.   Temperature dependence of the heat capacity of spins in a magnetic field with different values of S.

3.14   Phase transitions and the mean-field approximation

As explained in Chapter 1, with changing thermodynamic parameters such as temperature and pressure, the system can undergo a
phase transition. In the first-order phase transitions, the chemical
potentials μ of the two competing phases become equal at the phase
transition line, while on each side of this line, they are unequal and
only one of the phases is thermodynamically stable. The first-order
transition switches between two different phases (as, for instance,
between water and ice) and is thus abrupt. In contrast, second-order
transitions are gradual: The so-called order parameter continuously
grows from zero as the phase-transition line is crossed. Thermodynamic quantities are singular at second-order transitions.

   Phase transitions are complicated phenomena that arise due to
the interaction between particles in many-particle systems in the
thermodynamic limit $N \rightarrow \infty$. It can be shown that thermodynamic quantities of finite systems are nonsingular and, in particular, there are no second-order phase transitions. Up to now, in this
course, we studied systems without interaction. Inclusion of the latter
requires an extension of the formalism of statistical mechanics, which
is covered in Section 3.17. In such an extended formalism, one has
to calculate partition functions over the energy levels of the whole

system that, in general, depend on the interaction. In some cases, these energy levels are known exactly, but they are parametrized by many quantum numbers over which summation has to be carried out. In most cases, however, the energy levels are not known analytically, and their calculation is a huge quantum-mechanical problem. In both cases, the calculation of the partition function is a very difficult task. There are some models for which thermodynamic quantities have been calculated exactly, including models that possess second-order phase transitions. For other models, approximate numerical methods have been developed, which allow calculating phase transition temperatures and critical indices. It was shown that there are no phase transitions in 1D systems with short-range interactions. Also, there is no phase transition in the 2D Heisenberg model (Mermin–Wagner theorem).

It is most convenient to illustrate the physics of phase transitions by considering magnetic systems or the systems of spins introduced in Section 3.13. The simplest forms of interaction between different spins are Ising interaction $-J_{ij}\hat{S}_{iz}\hat{S}_{jz}$, including only z-components of the spins, and Heisenberg interaction $-J_{ij}\hat{\mathbf{S}}_i \cdot \hat{\mathbf{S}}_j$. The *exchange interaction* J_{ij} follows from the quantum mechanics of atoms and is of electrostatic origin. In most cases, practically, there is only interaction J between spins of the neighboring atoms in a crystal lattice because J_{ij} decreases exponentially with distance. The Hamiltonian of the whole system can be written as

$$\hat{H} = -\frac{1}{2}\sum_{ij} J_{ij}\hat{\mathbf{S}}_i \cdot \hat{\mathbf{S}}_j - g\mu_B\mathbf{B} \cdot \sum_i \hat{\mathbf{S}}_i \qquad (3.14.1)$$

for the Heisenberg model and similarly for the Ising model. The factor $1/2$ is introduced to compensate for the double counting of the same bond energy, say, $i = 1$, $j = 2$ and $i = 2$, $j = 1$. The energy of the interaction of spins with a uniform magnetic field $\mathbf{B}$, the so-called Zeeman interaction, is also taken into account, as in the preceding section.

For the Ising model, if the magnetic field is directed along the z-axis, all energy levels of the whole system are known exactly in terms of the quantum numbers m for each spin:

$$E_{\{m_i\}} = -\frac{1}{2}\sum_{ij} J_{ij}m_i m_j - g\mu_B B\sum_i m_i. \qquad (3.14.2)$$

Still, it is a problem to calculate the partition function of the Ising model, which is a sum over all m_i. It can be done relatively easily for a chain of spins $1/2$ with nearest-neighbor (nn) interaction. However, the solution is trivial and does not show any phase transition. For spins $1/2$ on a simple square lattice with nn interactions, the solution is rather complicated. It was found by the Swedish physicist Lars Onsager and shows a second-order magnetic phase transition. In three dimensions, there is no analytical solution.

For the Heisenberg model, energy levels are not known. In the case, $J > 0$, neighboring spins have the lowest interaction energy when they are collinear, and the state with all spins pointing in the same direction is the exact ground state of the system. For $B = 0$, this ground state has a continuous degeneracy because spins can be pointing in any direction. For the Ising model, all spins in the ground state are parallel or antiparallel with the z-axis, i.e., it is double-degenerate. In both cases, Ising and Heisenberg, the ground state for $J > 0$ is *ferromagnetic*.

The nature of the interaction between spins considered above suggests that at $T \to 0$, the system should fall into its ground state, so that thermodynamic averages of all spins approach their maximal values, $|\langle \hat{\mathbf{S}}_i \rangle| \to S$. With increasing temperature, excited levels of the system become populated and the average spin value decreases, $|\langle \hat{\mathbf{S}}_i \rangle| < S$. At high temperatures, all energy levels become populated, so that neighboring spins can have all possible orientations with respect to each other. In this case, if there is no external magnetic field acting on the spins (see Section 3.13), the average spin value should be exactly zero because there are as many spins pointing in one direction as there are spins pointing in any other direction. This is why the high-temperature state is called the *symmetric state*. If the temperature is now lowered, there should be a phase transition temperature, T_C (the Curie temperature for ferromagnets), below which the order parameter (average spin value) becomes nonzero. Below T_C, the symmetry of the state is *spontaneously* broken. This state is called the *ordered state*. Note that if there is an applied magnetic field, it will break the symmetry by creating a nonzero spin average at all temperatures, so that there is no sharp phase transition, only a gradual increase in $|\langle \hat{\mathbf{S}}_i \rangle|$ as the temperature is lowered.

Although the scenario outlined above looks persuasive, there are subtle effects that may preclude ordering and ensure $|\langle \hat{\mathbf{S}}_i \rangle| = 0$ at all nonzero temperatures. As said above, ordering does not occur in one dimension (spin chains) for both Ising and Heisenberg models. In two dimensions, the Ising model shows a phase transition. However, the 2D Heisenberg model does not order. In three dimensions, both Ising and Heisenberg models show a phase transition, and no analytical solution is available.

3.14.1 The mean-field approximation for ferromagnets

While a rigorous solution of the problem of phase transition is rather difficult, there is a simple approximate solution that captures the physics of phase transitions as described above and provides qualitatively correct results in most cases, including 3D systems. The idea of this approximate solution is to reduce the original many-spin problem to an effective self-consistent one-spin problem by considering one spin (say i) and replacing other spins in the interaction part of the Hamiltonian by their average thermodynamic values, $-J_{ij}\hat{\mathbf{S}}_i \cdot \hat{\mathbf{S}}_j \Rightarrow -J_{ij}\hat{\mathbf{S}}_i \cdot \langle\hat{\mathbf{S}}_j\rangle$. Further, one can consider the uniform system in which all spin averages are the same, $\langle\hat{\mathbf{S}}_j\rangle = \langle\hat{\mathbf{S}}\rangle$. This approach is called mean-field approximation (MFA) or molecular-field approximation, and it was first suggested by Weiss for ferromagnets. To work out this idea, one can define

$$\hat{\mathbf{S}}_i = \langle\hat{\mathbf{S}}\rangle + \delta\hat{\mathbf{S}}_i, \qquad (3.14.3)$$

where $\delta\hat{\mathbf{S}}_i$ is assumed to be small in comparison to $\langle\hat{\mathbf{S}}\rangle$. The exchange part of the Hamiltonian, Eq. (3.14.1), can be expanded up to the terms linear in $\delta\hat{\mathbf{S}}_i$ as follows:

$$\hat{H}_{\text{ex}} = -\frac{1}{2}\sum_{ij}J_{ij}(\langle\hat{\mathbf{S}}\rangle + \delta\hat{\mathbf{S}}_i) \cdot (\langle\hat{\mathbf{S}}\rangle + \delta\hat{\mathbf{S}}_j)$$

$$= -\frac{N}{2}Jz\langle\hat{\mathbf{S}}\rangle^2 - Jz\langle\hat{\mathbf{S}}\rangle \cdot \sum_i \delta\hat{\mathbf{S}}_i. \qquad (3.14.4)$$

Here, N is the number of spins in the system, and we assume the nn interactions, $\sum_j J_{ij} = Jz$, where z is the number of nearest neighbors for a spin in the lattice ($z = 6$ for the simple cubic lattice). Note that the factor $1/2$ was killed since there are two similar terms linear in $\delta\hat{\mathbf{S}}$. Then, substituting $\delta\hat{\mathbf{S}}_i = \hat{\mathbf{S}}_i - \langle\hat{\mathbf{S}}\rangle$, one can rewrite the exchange interaction as

$$\hat{H}_{\text{ex}} = -Jz\langle\hat{\mathbf{S}}\rangle \cdot \sum_i \hat{\mathbf{S}}_i + \frac{N}{2}Jz\langle\hat{\mathbf{S}}\rangle^2. \tag{3.14.5}$$

This is the Hamiltonian of a system of N-independent spins similar to that considered in the preceding section. Taking into account the Zeeman interaction, one can use the effective one-spin Hamiltonian similar to Eq. (3.13.7) that has the form

$$\hat{H} = -\hat{\mathbf{S}}\cdot(g\mu_B\mathbf{B} + Jz\langle\hat{\mathbf{S}}\rangle) + \frac{1}{2}Jz\langle\hat{\mathbf{S}}\rangle^2, \tag{3.14.6}$$

Now, one can use the formalism of Section (3.13) to solve the problem analytically. For the Heisenberg model, in the presence of a whatever weak applied field $\mathbf{B}$, the ordering will occur in the direction of $\mathbf{B}$. Choosing the z-axis in this direction, one can simplify Eq. (3.14.6) to

$$\hat{H} = -(g\mu_B B + Jz\langle\hat{S}_z\rangle)\hat{S}_z + \frac{1}{2}Jz\langle\hat{S}_z\rangle^2. \tag{3.14.7}$$

The same mean-field Hamiltonian is also valid for the Ising model if the magnetic field is applied along the z-axis. Thus, within the MFA (but not in general!), Ising and Heisenberg models are equivalent. The energy levels corresponding to Eq. (3.14.7) are

$$\varepsilon_m = -(g\mu_B B + Jz\langle\hat{S}_z\rangle)m + \frac{1}{2}Jz\langle\hat{S}_z\rangle^2,$$

$$m = -S, -S + 1, \ldots, S - 1, S. \tag{3.14.8}$$

At this point, one can just use the results of the preceding section and write down the expression for the average spin similar to Eq. (3.13.17): $\langle\hat{S}_z\rangle = b_S(y)$, where, in our case,

$$y \equiv \frac{g\mu_B B + Jz\langle\hat{S}_z\rangle}{k_B T}. \tag{3.14.9}$$

This yields the transcendental Curie–Weiss equation

$$\langle\hat{S}_z\rangle = b_S\left(\frac{g\mu_B B + Jz\langle\hat{S}_z\rangle}{k_B T}\right), \tag{3.14.10}$$

which defines $\langle\hat{S}_z\rangle$ in a self-consistent way.

In a more technical way, one can first calculate the partition function Z, Eq. (3.13.9), and then the free energy per spin:

$$F = -k_B T \ln Z = \frac{1}{2} Jz\langle \hat{S}_z \rangle^2 - k_B T \ln \frac{\sinh\left[(S + 1/2)y\right]}{\sinh(y/2)}. \quad (3.14.11)$$

To find the actual value of the order parameter $\langle \hat{S}_z \rangle$ at any temperature T and magnetic field B, one can minimize F with respect to $\langle \hat{S}_z \rangle$, as explained in Chapter 1. This yields the Curie–Weiss equation (3.14.10):

$$0 = \frac{\partial F}{\partial \langle \hat{S}_z \rangle} = Jz\langle \hat{S}_z \rangle - Jzb_S(y). \quad (3.14.12)$$

For $B = 0$, Eq. (3.14.10) has the only solution $\langle \hat{S}_z \rangle = 0$ at high temperatures. As $b_S(y)$ has the maximal slope at $y = 0$, it is sufficient that this slope (with respect to $\langle \hat{S}_z \rangle$) becomes smaller than 1 to exclude any solution other than $\langle \hat{S}_z \rangle = 0$. Using Eq. (3.13.22), one obtains that the only solution $\langle \hat{S}_z \rangle = 0$ is realized for $T \geq T_C$, where

$$T_C = \frac{S(S + 1)}{3} \frac{Jz}{k_B} \quad (3.14.13)$$

is the Curie temperature within the MFA. Below T_C, the slope of $b_S(y)$ with respect to $\langle \hat{S}_z \rangle$ exceeds 1, so that there are three solutions for $\langle \hat{S}_z \rangle$: One solution $\langle \hat{S}_z \rangle = 0$ and two symmetric solutions $\langle \hat{S}_z \rangle \neq 0$ (see Fig. 3.7). The latter corresponds to the lower free energy than the solution $\langle \hat{S}_z \rangle = 0$, thus they are thermodynamically stable (see Fig. 3.8). These solutions describe the ordered state below T_C.

Slightly below T_C, the value of $\langle \hat{S}_z \rangle$ is still small and can be found by expanding $b_S(y)$ up to y^3. In particular, for $S = 1/2$, one has

$$b_{1/2}(y) = \frac{1}{2} \tanh \frac{y}{2} \cong \frac{1}{4} y - \frac{1}{48} y^3. \quad (3.14.14)$$

Using $T_C = (1/4)Jz/k_B$ and thus $Jz = 4k_B T_C$ for $S = 1/2$, one can rewrite Eq. (3.14.10) with $B = 0$ in the form

$$\langle \hat{S}_z \rangle = \frac{T_C}{T} \langle \hat{S}_z \rangle - \frac{4}{3} \left(\frac{T_C}{T}\right)^3 \langle \hat{S}_z \rangle^3. \quad (3.14.15)$$

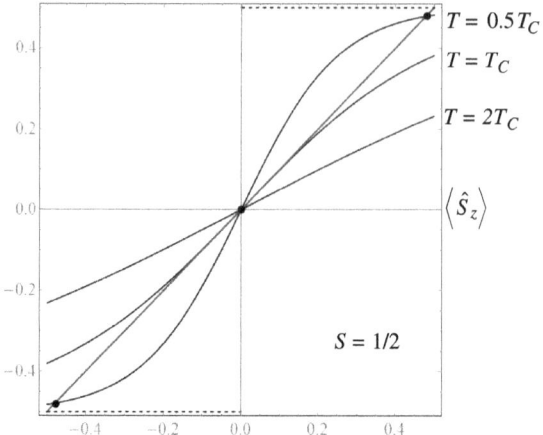

Figure 3.7. Graphic solution of the Curie–Weiss equation.

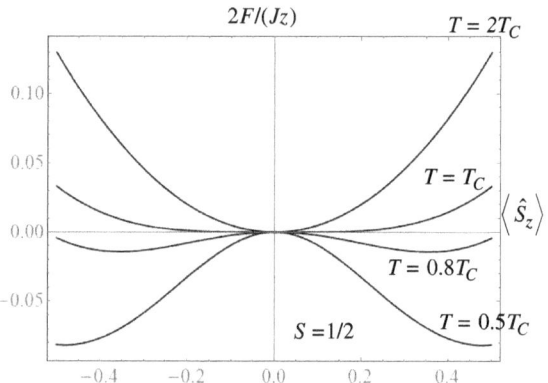

Figure 3.8. Free energy of a ferromagnet vs $\langle \hat{S}_z \rangle$ within the MFA (arbitrary vertical shift).

One of the solutions is $\langle \hat{S}_z \rangle = 0$, while the other two solutions are

$$\frac{\langle \hat{S}_z \rangle}{S} = \pm \frac{T}{T_C} \sqrt{3 \left(1 - \frac{T}{T_C} \right)}, \quad S = \frac{1}{2}. \qquad (3.14.16)$$

As this result is only valid for T near T_C, one can discard the factor in front of the square root and simplify this formula to

$$\frac{\langle \hat{S}_z \rangle}{S} = \pm \sqrt{3 \left(1 - \frac{T}{T_C} \right)}, \quad S = \frac{1}{2}. \qquad (3.14.17)$$

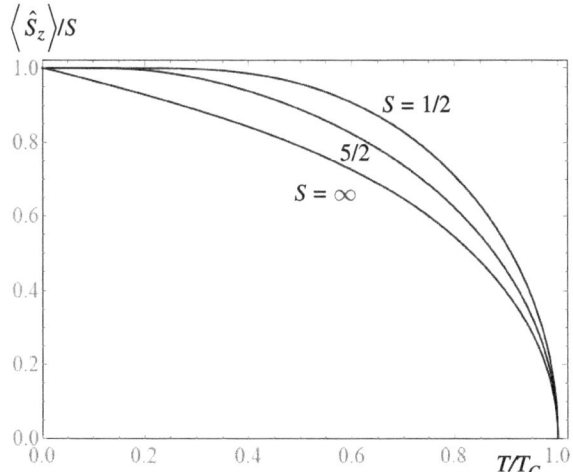

Figure 3.9. Temperature dependences of normalized spin averages $\langle \hat{S}_z \rangle / S$ for $S = 1/2, 5/2, \infty$ obtained from the numerical solution of the Curie–Weiss equation.

Although obtained near T_C, the result in this form is only by the factor $\sqrt{3}$ off at $T = 0$. The singularity of the order parameter near T_C is square root, so that for the magnetization critical index defined as

$$\langle \hat{S}_z \rangle \propto M \propto (T_C - T)^{\beta}, \qquad (3.14.18)$$

one has $\beta = 1/2$. Results of the numerical solution of the Curie–Weiss equation with $B = 0$ for different S are shown in Fig. 3.9. Note that the magnetization is related to the spin average by Eq. (3.13.25).

Let us now consider the magnetic susceptibility per spin χ defined by Eq. (3.13.28) above T_C. Linearizing Eq. (3.14.10) with the use of Eq. (3.13.22), one obtains

$$\langle \hat{S}_z \rangle = \frac{S(S+1)}{3} \frac{g\mu_B B + Jz\langle \hat{S}_z \rangle}{k_B T} = \frac{S(S+1)}{3} \frac{g\mu_B B}{k_B T} + \frac{T_C}{T} \langle \hat{S}_z \rangle.$$

$$(3.14.19)$$

The solution of this equation reads

$$\langle \hat{S}_z \rangle = \frac{S(S+1)}{3} \frac{g\mu_B B}{k_B T} \frac{1}{1 - T_C/T} = \frac{S(S+1)}{3} \frac{g\mu_B B}{k_B(T_C - T)}.$$

$$(3.14.20)$$

Then the magnetic susceptibility can be calculated by differentiation:

$$\chi = \frac{\partial \langle \mu_z \rangle}{\partial B} = g\mu_B \frac{\partial \langle \hat{S}_z \rangle}{\partial B} = \frac{S(S+1)}{3} \frac{(g\mu_B)^2}{k_B (T - T_C)}. \qquad (3.14.21)$$

In contrast to noninteracting spins, Eq. (3.13.32), the susceptibility diverges at $T = T_C$ rather than at $T = 0$. The inverse susceptibility χ^{-1} is a straight line crossing the T-axis at T_C. In the theory of phase transitions, the critical index for the susceptibility γ is defined as

$$\chi \propto (T - T_C)^{-\gamma}. \qquad (3.14.22)$$

One can see that $\gamma = 1$ within the MFA.

At $T = T_C$, there is no linear susceptibility, and the magnetization is a singular function of the magnetic field that defines another critical index, δ. In this case, the Curie–Weiss equation can be expanded for small $\langle \hat{S}_z \rangle$ and small B. Using Eq. (3.14.14), at $T = T_C$, one obtains

$$0 = \frac{g\mu_B B}{4k_B T} - \frac{4}{3} \langle \hat{S}_z \rangle^3, \qquad (3.14.23)$$

c.f. Eq. (3.14.15). Thus,

$$\langle \hat{S}_z \rangle = \left(\frac{3g\mu_B B}{16 k_B T} \right)^{1/3} \qquad S = \frac{1}{2}. \qquad (3.14.24)$$

The critical isotherm index δ is defined by

$$\langle \hat{S}_z \rangle \propto B^{1/\delta}, \qquad (3.14.25)$$

thus within the MFA, $\delta = 3$.

More precise methods yield smaller values of β and larger values of γ and δ, depending on the details of the model, such as the number of interacting spin components (1 for Ising and 3 for Heisenberg) and the dimension of the lattice. Critical indices are insensitive to factors such as the lattice structure and the value of the spin S. On the other hand, the value of T_C does not possess such *universality* and depends on all parameters of the problem. In 3D, accurate values of T_C are by up to 25% lower than their mean-field values.

3.15 1D Ising model

The 1D Ising model is a chain of magnetic atoms with spins S interacting with their nearest neighbors in the chain via the coupling of their z-components. The Hamiltonian of the 1D Ising model reads

$$\hat{H} = -J \sum_{i=1}^{N-1} \hat{S}_{iz}\hat{S}_{i+1,z} - g\mu_B B \sum_{i=1}^{N} \hat{S}_{iz}, \qquad (3.15.1)$$

where N is the number of atoms in the chain, the exchange coupling $J > 0$ for the ferromagnetic interaction, and B is the magnetic field applied along the z-axis. The quantum energy levels of the Ising model are trivial and correspond to the values of $\hat{S}_{iz}$ equal to m_i taking $2S+1$ values between $-S$ and S. (In the presence of a transverse field or a coupling of other spin components, the quantum problem becomes nontrivial.) Thus, the energy of the system divided by the temperature that enters the partition function can be written as

$$\beta E \equiv \frac{E}{k_B T} = -\theta \sum_{i=1}^{N-1} m_i m_{i+1} - \rho \sum_{i=1}^{N} m_i, \qquad (3.15.2)$$

where we have introduced the dimensionless exchange and field parameters

$$\theta \equiv \beta J = \frac{J}{k_B T}, \qquad \rho \equiv \frac{g\mu_B B}{k_B T}. \qquad (3.15.3)$$

One can calculate the partition function analytically if one considers the ring instead of the open chain. In the thermodynamic limit, there should be no difference between the two model variants, but for the ring, the analytical calculation exists for any N. Thus, we add the term $-\theta m_N m_1$ to βE (the last spin is coupled with the first one). The partition function for the Ising ring is given by

$$Z = \sum_{m_1,m_2,\ldots,m_N} \exp(\theta m_1 m_2 + \rho m_1 + \theta m_2 m_3$$
$$+ \rho m_2 + \cdots + \theta m_N m_1 + \rho m_N). \qquad (3.15.4)$$

It is convenient to make a trick that allows representing the partition function as a product of matrices. For this, we redistribute the field terms:

$$Z = \sum_{m_1,m_2,\ldots,m_N} \exp\left[\theta m_1 m_2 + \frac{\rho}{2}(m_1 + m_2) + \theta m_2 m_3\right.$$

$$\left. + \frac{\rho}{2}(m_2 + m_3) + \cdots + \theta m_N m_1 + \frac{\rho}{2}(m_N + m_1)\right]. \qquad (3.15.5)$$

Then,

$$Z = \sum_{m_1,m_2,\ldots,m_N} A_{m_1 m_2} A_{m_2 m_3} \cdots A_{m_{N-1} m_N} A_{m_N m_1}$$

$$= \sum_{m_1} \left(\mathbb{A}^N\right)_{m_1 m_1} = \operatorname{Tr}\left(\mathbb{A}^N\right) \qquad (3.15.6)$$

is the trace of the Nth power of the $(2S + 1) \times (2S + 1)$ matrix with matrix elements

$$A_{mk} = \exp\left(\theta mk + \frac{\rho}{2}(m + k)\right), \quad m, k = -S, -S + 1, \ldots, S. \qquad (3.15.7)$$

For $S = 1/2$, this is a 2×2 matrix:

$$\mathbb{A} = \begin{pmatrix} \exp(\theta/4 - \rho/2) & \exp(-\theta/4) \\ \exp(-\theta/4) & \exp(\theta/4 + \rho/2) \end{pmatrix}. \qquad (3.15.8)$$

It can be easily shown that the trace of a matrix is invariant under the unitary transformation

$$\mathbb{B} = \mathbb{U}^{-1}\mathbb{A}\mathbb{U}, \qquad (3.15.9)$$

where $\mathbb{U}$ is a unitary matrix and the inverse unitary matrix $\mathbb{U}^{-1}$ is just the transposed and complex conjugate of $\mathbb{U}$. As one can make a cyclic permutation of matrices under the Tr operator, one can write

$$\operatorname{Tr}\left(\mathbb{A}^N\right) = \operatorname{Tr}\left(\mathbb{A}^N \mathbb{U}\mathbb{U}^{-1}\right) = \operatorname{Tr}\left(\mathbb{U}^{-1}\mathbb{A}^N \mathbb{U}\right)$$

$$= \operatorname{Tr}\left(\mathbb{U}^{-1}\mathbb{A}\mathbb{U}\mathbb{U}^{-1}\mathbb{A}\mathbb{U}\ldots\mathbb{U}^{-1}\mathbb{A}\mathbb{U}\right) = \operatorname{Tr}\left(\mathbb{B}^N\right), \qquad (3.15.10)$$

which proves the statement above. One can choose the unitary transformation that diagonalizes $\mathbb{A}$, in this case, for $S = 1/2$,

$$\mathbb{B} = \begin{pmatrix} \lambda_1 & 0 \\ 0 & \lambda_2 \end{pmatrix}, \quad \mathbb{B}^N = \begin{pmatrix} \lambda_1^N & 0 \\ 0 & \lambda_2^N \end{pmatrix}. \tag{3.15.11}$$

Thus, the problem is reduced to calculating the eigenvalues of $\mathbb{A}$ above. The secular equation has the form

$$0 = \begin{vmatrix} \exp(\theta/4 - \rho/2) - \lambda & \exp(-\theta/4) \\ \exp(-\theta/4) & \exp(\theta/4 + \rho/2) - \lambda \end{vmatrix}$$
$$= [\exp(\theta/4 - \rho/2) - \lambda][\exp(\theta/4 + \rho/2) - \lambda] - \exp(-\theta/2) \tag{3.15.12}$$

or

$$\lambda^2 - e^{\theta/4}(e^{\rho/2} + e^{-\rho/2})\lambda + e^{\theta/2} - e^{-\theta/2} = 0 \tag{3.15.13}$$

or, finally,

$$\lambda^2 - 2e^{\theta/4}\cosh\frac{\rho}{2}\lambda + 2\sinh\frac{\theta}{2} = 0. \tag{3.15.14}$$

The solution of this quadratic equation is

$$\lambda_{1,2} = e^{\theta/4}\cosh\frac{\rho}{2} \pm \sqrt{e^{\theta/2}\cosh^2\frac{\rho}{2} - 2\sinh\frac{\theta}{2}}$$
$$= e^{\theta/4}\cosh\frac{\rho}{2} \pm \sqrt{e^{\theta/2}\cosh^2\frac{\rho}{2} - e^{\theta/2} + e^{-\theta/2}}$$
$$= e^{\theta/4}\left(\cosh\frac{\rho}{2} \pm \sqrt{\sinh^2\frac{\rho}{2} + e^{-\theta}}\right). \tag{3.15.15}$$

Now, the partition function is

$$Z = \lambda_1^N + \lambda_2^N \Rightarrow e^{N\theta/4}\left(\cosh\frac{\rho}{2} + \sqrt{\sinh^2\frac{\rho}{2} + e^{-\theta}}\right)^N, \tag{3.15.16}$$

where we have discarded the smaller eigenvalue that makes a negligible contribution for large N. One can see that Z does not have any singularities, so that in the 1D Ising model, there are no phase transitions.

The 1D Ising model with an arbitrary S can be solved in a similar way by diagonalizing a $(2S + 1) \times (2S + 1)$ matrix. However, this solution is cumbersome and can be done only numerically in general. There is no phase transition for any S, including $S \to \infty$.

All thermodynamic quantities can be obtained from Z. Let us first calculate the internal energy in zero field. In this case, $\rho = 0$ and

$$Z = e^{N\theta/4}(1 + e^{-\theta/2})^N = (e^{\theta/4} + e^{-\theta/4})^N = 2^N \cosh^N \frac{\theta}{4}.$$

$$(3.15.17)$$

The energy is given by

$$U = -\frac{\partial \ln Z}{\partial \beta} = -J\frac{\partial \ln Z}{\partial \theta} = -NJ\frac{\partial \ln \cosh \frac{\theta}{4}}{\partial \theta}$$

$$= -\frac{1}{4}NJ\tanh\frac{\theta}{4} = -\frac{1}{4}NJ\tanh\frac{J}{4k_BT}.$$

$$(3.15.18)$$

In the limit, $T \to 0$ ($\theta \to \infty$), one has

$$U = -\frac{1}{4}NJ,$$

$$(3.15.19)$$

which can be obtained directly as the ground state of the Hamiltonian, Eq. (3.15.1). In the opposite limit, $k_BT \gg J$ ($\theta \ll 1$), the energy behaves as

$$U \cong -\frac{1}{16}NJ\theta = -\frac{1}{16}N\frac{J^2}{k_BT}.$$

$$(3.15.20)$$

The heat capacity is given by

$$C = \frac{\partial U}{\partial T} = Nk_B\left(\frac{J}{4k_BT}\right)^2/\cosh^2\frac{J}{4k_BT}.$$

$$(3.15.21)$$

This is exponentially small at $k_BT \ll J$ and behaves as

$$C = Nk_B\left(\frac{J}{4k_BT}\right)^2$$

$$(3.15.22)$$

for $k_BT \gg J$. At $k_BT \sim J$, the heat capacity has a broad maximum.

The magnetization of the Ising model per spin is given by the standard formula, Eq. (3.13.24). The spin polarization $\langle \hat{S}_z \rangle$ is defined by

$$
\langle \hat{S}_z \rangle = \frac{1}{N} \left\langle \sum_{i=1}^{N} m_i \right\rangle = \frac{1}{N} \frac{1}{Z} \sum_{m_1, m_2, \ldots, m_N} \left(\sum_{i=1}^{N} m_i \right)
$$
$$
\times \exp \left(\theta m_1 m_2 + \theta m_2 m_3 + \cdots + \theta m_N m_1 + \rho \sum_{i=1}^{N} m_i \right),
$$
$$(3.15.23)$$

which can be expressed as

$$
\langle \hat{S}_z \rangle = \frac{1}{N} \frac{1}{Z} \frac{\partial Z}{\partial \rho} = \frac{1}{N} \frac{\partial \ln Z}{\partial \rho}. \tag{3.15.24}
$$

From Eq. (3.15.16), one obtains

$$
\langle \hat{S}_z \rangle = \frac{\partial}{\partial \rho} \ln \left(\cosh \frac{\rho}{2} + \sqrt{\sinh^2 \frac{\rho}{2} + e^{-\theta}} \right)
$$
$$
= \frac{\frac{1}{2} \sinh \frac{\rho}{2} + \frac{1}{2} \frac{\sinh \frac{\rho}{2} \cosh \frac{\rho}{2}}{\sqrt{\sinh^2 \frac{\rho}{2} + e^{-\theta}}}}{\cosh \frac{\rho}{2} + \sqrt{\sinh^2 \frac{\rho}{2} + e^{-\theta}}}, \tag{3.15.25}
$$

which simplifies to

$$
\langle \hat{S}_z \rangle = \frac{1}{2} \frac{\sinh \frac{\rho}{2}}{\sqrt{\sinh^2 \frac{\rho}{2} + e^{-\theta}}}. \tag{3.15.26}
$$

One can see that the spin polarization vanishes in the limit $B \to 0$ ($\rho \to 0$), thus there is no spontaneous ordering in this model. In the high-field limit, $\rho \to \infty$, one has $\langle \hat{S}_z \rangle = 1/2$, the full alignment of the spins. The magnetic susceptibility per spin is given by

$$
\chi = \frac{\partial \langle \mu_z \rangle}{\partial B} = g\mu_B \frac{\partial \langle \hat{S}_z \rangle}{\partial B} = \frac{(g\mu_B)^2}{k_B T} \frac{\partial \langle \hat{S}_z \rangle}{\partial \rho}. \tag{3.15.27}
$$

Using the formula above, one obtains

$$\chi = \frac{(g\mu_B)^2}{k_BT}\frac{1}{4}\left(\frac{\cosh\frac{\rho}{2}}{\sqrt{\sinh^2\frac{\rho}{2}+e^{-\theta}}} - \frac{\sinh\frac{\rho}{2}\cosh\frac{\rho}{2}}{\left(\sinh^2\frac{\rho}{2}+e^{-\theta}\right)^{3/2}}\right)$$

$$= \frac{(g\mu_B)^2}{k_BT}\frac{\cosh\frac{\rho}{2}}{4\left(\sinh^2\frac{\rho}{2}+e^{-\theta}\right)^{3/2}}\left[\sinh^2\frac{\rho}{2}+e^{-\theta}-\sinh\frac{\rho}{2}\right].$$

$$(3.15.28)$$

In zero field, $\rho = 0$, this becomes

$$\chi = \frac{(g\mu_B)^2}{4k_BT}e^{\theta/2} = \frac{(g\mu_B)^2}{4k_BT}\exp\frac{J}{2k_BT}. \qquad (3.15.29)$$

One can see that the magnetic susceptibility exponentially diverges at $T = 0$, which is the most interesting result for the 1D Ising model. At high temperatures, the exponential term tends to 1, and the result is the susceptibility of noninteracting spins, a particular case of Eq. (3.13.32) for $S = 1/2$.

3.16   Problems

3.16.1   Magnetic susceptibility below T_C in the mean-field approximation

Calculate the zero-field magnetic susceptibility per spin below T_C in the MFA. Work out its form close to T_C.

Solution: The spin polarization $\langle \hat{S}_z \rangle$ in the MFA satisfies the Curie–Weiss equation

$$\langle \hat{S}_z \rangle = b_S\left(\frac{g\mu_B B + Jz\langle\hat{S}_z\rangle}{k_BT}\right). \qquad (3.16.1)$$

The magnetic susceptibility per spin is defined as

$$\chi = \frac{\partial\langle\mu_z\rangle}{\partial B} = g\mu_B\frac{\partial\langle\hat{S}_z\rangle}{\partial B}. \qquad (3.16.2)$$

Using Eq. (3.16.1), one can write

$$\frac{\partial \langle \hat{S}_z \rangle}{\partial B} = b'_S \left(\frac{g\mu_B B + Jz\langle \hat{S}_z \rangle}{k_B T} \right) \left[\frac{g\mu_B}{k_B T} + \frac{Jz}{k_B T} \frac{\partial \langle \hat{S}_z \rangle}{\partial B} \right]. \quad (3.16.3)$$

This is an equation for the derivative that has the solution

$$\frac{\partial \langle \hat{S}_z \rangle}{\partial B} = \frac{g\mu_B}{k_B T} \frac{b'_S}{1 - b'_S \frac{Jz}{k_B T}}, \quad (3.16.4)$$

where the argument of b'_S is suppressed for brevity. It is convenient to express Jz via T_C using

$$T_C = \frac{S(S+1)}{3} \frac{Jz}{k_B}. \quad (3.16.5)$$

This yields

$$\frac{\partial \langle \hat{S}_z \rangle}{\partial B} = \frac{g\mu_B}{k_B T} \frac{b'_S}{1 - \frac{3b'_S}{S(S+1)} \frac{T_C}{T}}. \quad (3.16.6)$$

This is the solution of the problem at all temperatures. One has to solve the Curie–Weiss equation numerically to find $\langle \hat{S}_z \rangle$ and substitute the result into the argument of b'_S.

The most important is the zero-field susceptibility, $B = 0$. In this case, above T_C, one has $\langle \hat{S}_z \rangle = 0$, so that b'_S has zero argument, $b'_S(0) = S(S+1)/3$ and one obtains

$$\chi - \frac{\partial \langle \mu_z \rangle}{\partial B} - g\mu_B \frac{\partial \langle \hat{S}_z \rangle}{\partial B} - \frac{S(S+1)}{3} \frac{(g\mu_B)^2}{k_B T} \frac{1}{1 - \frac{T_C}{T}}$$

$$= \frac{S(S+1)}{3} \frac{(g\mu_B)^2}{k_B (T - T_C)}. \quad (3.16.7)$$

Below T_C, there is a nonzero spin polarization, and one has to do some work to find $\langle \hat{S}_z \rangle$ just below T_C and substitute it into b'_S.

The spin polarization for $S = 1/2$ has been found in the lecture for $S = 1/2$. Here, we need to do it in general using the expansion

$$b_S(y) = \frac{1}{3}S(S+1)y - \frac{1}{90}S(S+1)(2S^2 + 2S + 1)y^3 + \cdots \quad (3.16.8)$$

Then, near T_C, the Curie–Weiss equation with $B = 0$ becomes

$$\langle \hat{S}_z \rangle = \frac{1}{3}S(S+1)\frac{Jz\langle \hat{S}_z \rangle}{k_B T} - \frac{1}{90}S(S+1)(2S^2 + 2S + 1)\left(\frac{Jz\langle \hat{S}_z \rangle}{k_B T}\right)^3.$$

$$(3.16.9)$$

Canceling $\langle \hat{S}_z \rangle$ and expressing Jz via T_C, one simplifies this equation to

$$1 = \frac{T_C}{T} - \frac{3(2S^2 + 2S + 1)}{10\left[S(S+1)\right]^2}\left(\frac{T_C}{T}\right)^3 \langle \hat{S}_z \rangle^2, \quad (3.16.10)$$

which defines $\langle \hat{S}_z \rangle$ below T_C. It is convenient to write the solution as

$$\left(\frac{\langle \hat{S}_z \rangle}{S}\right)^2 = \left(\frac{T}{T_C}\right)^3 \frac{10(S+1)^2}{3(2S^2 + 2S + 1)}\left(\frac{T_C}{T} - 1\right)$$

$$= \left(\frac{T}{T_C}\right)^2 \frac{10(S+1)^2}{3(2S^2 + 2S + 1)}\left(1 - \frac{T}{T_C}\right) \quad (3.16.11)$$

to embrace the case of $S \to \infty$. The formula for the spontaneous spin polarization reads

$$\frac{\langle \hat{S}_z \rangle}{S} = \sqrt{\frac{10(S+1)^2}{3(2S^2 + 2S + 1)}\left(1 - \frac{T}{T_C}\right)}, \quad (3.16.12)$$

which generalizes the $S = 1/2$ result obtained above for the general S. Here, also, the factor T/T_C has been discarded near the phase transition point.

For the derivative $b'_S(y)$, we use the expansion

$$b'_S(y) = \frac{1}{3}S(S+1) - \frac{1}{30}S(S+1)(2S^2 + 2S + 1)y^2 + \cdots, \quad (3.16.13)$$

which follows from that for $b_S(y)$. In Eq. (3.16.6), one obtains

$$\frac{3b'_S}{S(S+1)} = 1 - \frac{(2S^2 + 2S + 1)}{10} \left(\frac{Jz\langle\hat{S}_z\rangle}{k_BT}\right)^2$$

$$= 1 - \frac{(2S^2 + 2S + 1)}{10} \left(\frac{3}{S(S+1)}\right)^2 \left(\frac{T_C}{T}\right)^2 \langle\hat{S}_z\rangle^2$$

$$(3.16.14)$$

and, finally,

$$\frac{3b'_S}{S(S+1)} = 1 - \frac{9(2S^2 + 2S + 1)}{10(S+1)^2} \left(\frac{T_C}{T}\right)^2 \left(\frac{\langle\hat{S}_z\rangle}{S}\right)^2. \qquad (3.16.15)$$

Substituting here the solution for the spin polarization, Eq. (3.16.11), one obtains

$$\frac{3b'_S}{S(S+1)} = 1 - \frac{9(2S^2 + 2S + 1)}{10(S+1)^2} \left(\frac{T_C}{T}\right)^2 \left(\frac{T}{T_C}\right)^2$$

$$\times \frac{10(S+1)^2}{3(2S^2 + 2S + 1)} \left(1 - \frac{T}{T_C}\right) = 1 - 3\left(1 - \frac{T}{T_C}\right),$$

$$(3.16.16)$$

a big simplification! Now, substituting this into Eq. (3.16.6), one obtains

$$\chi = \frac{\partial\langle\mu_z\rangle}{\partial B} = g\mu_B \frac{\partial\langle\hat{S}_z\rangle}{\partial B} = \frac{S(S+1)}{3} \frac{(g\mu_B)^2}{k_BT} \frac{1}{1 - [1 - 3(1 - \frac{T}{T_C})]\frac{T_C}{T}}$$

$$(3.16.17)$$

and, simplifying the denominator,

$$\chi = \frac{S(S+1)}{3} \frac{(g\mu_B)^2}{k_BT} \frac{1}{1 - \frac{T_C}{T} + 3(\frac{T_C}{T} - 1)}$$

$$= \frac{S(S+1)}{3} \frac{(g\mu_B)^2}{k_BT} \frac{1}{2(\frac{T_C}{T} - 1)}. \qquad (3.16.18)$$

Finally, below T_C, the zero-field susceptibility is

$$\chi = \frac{S(S+1)\,(g\mu_B)^2}{6}\frac{1}{k_B}\frac{1}{T_C - T}. \tag{3.16.19}$$

The coefficient here is two times smaller than that in the susceptibility above T_C. The general result near T_C can be written as

$$\chi = \frac{S(S+1)\,(g\mu_B)^2}{3}\frac{1}{k_B}\frac{1}{|T_C - T|}\begin{cases}1, & T > T_C,\\ 1/2, & T < T_C.\end{cases} \tag{3.16.20}$$

3.16.2   Two interacting Ising spins

Consider the model of two coupled spins with the Hamiltonian

$$\hat{H} = -g\mu_B B\,(S_{1,z} + S_{2,z}) - J S_{1,z} S_{2,z},$$

where B is the external magnetic field and J is the so-called exchange interaction, ferromagnetic for $J > 0$ and antiferromagnetic for $J < 0$. The model above in which only the z-components of the spins are coupled is called the Ising model. The energy levels of this system are given by

$$\varepsilon_{m_1 m_2} = -g\mu_B B\,(m_1 + m_2) - J m_1 m_2,$$

where the quantum numbers take the values $-S \leq m_1, m_2 \leq S$. Write down the expression for the partition function of the system. Can it be calculated analytically for a general S? If not, perform the calculation for $S = 1/2$ only. Calculate the internal energy, heat capacity, magnetization induced by the magnetic field, and the magnetic susceptibility. Analyze ferromagnetic and antiferromagnetic cases.

Solution: The partition function of the system is given by

$$Z_S = \sum_{m_1,m_2=-S}^{S} e^{-\beta\varepsilon_{m_1 m_2}} = \sum_{m_1,m_2=-S}^{S} e^{\beta h(m_1+m_2)+\beta J m_1 m_2}, \tag{3.16.21}$$

where $h \equiv g\mu_B B$. For a general spin S, one can perform only one summation analytically. One can use the results for a single spin in a magnetic field and write

$$Z_S = \sum_{m_1=-S}^{S} e^{\beta h m_1} \sum_{m_2=-S}^{S} e^{\beta(h+Jm_1)m_2}$$

$$= \sum_{m=-S}^{S} e^{\beta h m} \frac{\sinh\left[(S+1/2)\beta\left(h+Jm\right)\right]}{\sinh\left[\beta\left(h+Jm\right)/2\right]}.$$

The remaining sum most probably cannot be calculated analytically. For $S = 1/2$, the expression above simplifies to

$$Z_{1/2} = \sum_{m=-1/2}^{1/2} e^{\beta h m} 2 \cosh\left[\frac{\beta\left(h+Jm\right)}{2}\right],$$

i.e.,

$$Z_{1/2} = 2\left\{e^{\beta h/2}\cosh\left[\frac{\beta\left(h+J/2\right)}{2}\right] + e^{-\beta h/2}\cosh\left[\frac{\beta\left(h-J/2\right)}{2}\right]\right\}.$$

This expression becomes

$$Z_{1/2} = 2\left[e^{\beta J/4}\cosh\left(\beta h\right) + e^{-\beta J/4}\right]. \tag{3.16.22}$$

Finally, in the zero field, the result simplifies to

$$Z_{1/2} = 4\cosh(\beta J/4).$$

Let us calculate the internal energy and heat capacity in the zero field. One obtains

$$U = -N\frac{\partial \ln Z}{\partial \beta} = -\frac{NJ}{4}\tanh\left(\frac{\beta J}{4}\right),$$

where N is the number of two-spin systems. In the limit of low temperatures, the hyperbolic tangent tends to 1, and one obtains the anticipated result $U = -NJ/4$ (the two coupled spins are parallel for $J > 0$). In the case of antiferromagnetic coupling, $J < 0$, one has $\tanh\left(\beta J/4\right) \to -1$ for $T \to 0$, so that for both ferromagnetic and antiferromagnetic coupling, one obtains $U = -N|J|/4$ at zero temperature. In general, as U is an even function of J, it is the same for the ferromagnetic and antiferromagnetic coupling.

The average spin value per two-spin system can be obtained by the differentiation of Eq. (3.16.21) and is given by

$$\langle S_z \rangle = \langle m_1 + m_2 \rangle = \frac{1}{\mathcal{Z}} \frac{\partial \mathcal{Z}}{\partial (\beta h)}. \qquad (3.16.23)$$

Using Eq. (3.16.22), one obtains

$$\langle S_z \rangle = \frac{e^{\beta J/4} \sinh (\beta h)}{e^{\beta J/4} \cosh (\beta h) + e^{-\beta J/4}} = \frac{\sinh (\beta h)}{\cosh (\beta h) + e^{-\beta J/2}}.$$

The susceptibility can be obtained by differentiating this expression with respect to the magnetic field,

$$\chi = \frac{\partial \langle \mu_z \rangle}{\partial B} = g \mu_B \frac{\partial \langle \hat{S}_z \rangle}{\partial B}.$$

In particular, at the zero field,

$$\chi = \frac{(g \mu_B)^2}{k_B T} \frac{1}{1 + e^{-\beta J/2}}.$$

In the ferromagnetic case, $J > 0$, the exponential is very small at low temperatures, $\beta J \gg 1$, so that the susceptibility has a regular value comparable with that of an isolated spin. On the contrary, in the antiferromagnetic case, $J < 0$, the exponential is large, and thus, the susceptibility is very small. Try to explain this in physical terms and draw the dependence $\langle S_z \rangle$ on the magnetic field for the antiferromagnetic coupling at low temperatures.

3.17   The grand canonical ensemble

In the previous chapters, we considered statistical properties of systems of noninteracting and *distinguishable* particles. Microstate (1,2) (particle 1 in state 1 and particle 2 in state 2) and microstate (2,1) (particle 2 in state 1 and particle 1 in state 2) were counted as different microstates within the same macrostate. However, quantum-mechanical particles such as electrons are indistinguishable from each other. Thus, microstates (1,2) and (2,1) should be counted as the same microstate, one particle in state 1 and one in state 2, without specifying which particle is in which state. Accordingly, the method based on the minimization of the Boltzmann entropy of a large system, Eq. (3.5.1), becomes invalid and has to be modified by removing double counting of the same microstates. This changes the foundations of statistical description of such systems.

In addition to the indistinguishability of quantum particles, it follows from the relativistic quantum theory that particles with half-integer spin, called *fermions*, cannot occupy any quantum state more than once. There is no place for more than one fermion in any quantum state, which is the so-called *exclusion principle*. An important example of fermions is the electron. On the contrary, particles with integer spin, called *bosons*, can be put into a quantum state in unlimited numbers. Examples of bosons are Helium and other atoms with an integer spin. Note that in the case of fermions, even for a large system, one cannot use the Stirling formula, Eq. (3.3.1), to simplify combinatorial expressions since the occupation numbers of quantum states can be only 0 and 1. This is an additional reason why the Boltzmann formalism does not work here.

Finally, the Boltzmann formalism of quantum statistics does not work for systems with interaction because the quantum states of such systems are quantum states of the system as a whole rather than one-particle quantum states. In the preceding section, we tacitly used an extension of the Boltzmann method for a system with interaction — the 1D Ising model. The idea was to use an ensemble of systems instead of an ensemble of noninteracting particles.

The idea of considering an ensemble of systems that will be worked out in the following can be justified by partitioning a large system into many still large subsystems. At least for short-range interactions, the energy of the interaction across the subsystems' boundaries is much smaller than the energy due to the interactions inside the subsystems and can be neglected. Then the subsystems become an ensemble of independent systems for which the same method can be applied as was applied to the ensemble of independent particles in the Boltzmann method. There are different variants of the formalism, from which we consider only one — the *grand canonical ensemble*.

The grand canonical ensemble consists of $\mathcal{N}$ identical systems, each possessing quantum energy levels ξ. The ensemble contains any number $\mathcal{N}_\xi$ of systems in states ξ. The total number of system in the ensemble is

$$\mathcal{N} = \sum_\xi \mathcal{N}_\xi. \tag{3.17.1}$$

The number of particles in the states ξ is not fixed, and each state ξ has N_ξ particles (do not confuse $\mathcal{N}_\xi$ with N_ξ). We require that the average number of particles in the system and the average energy of the system over the ensemble are fixed:

$$N = \frac{1}{\mathcal{N}} \sum_\xi N_\xi \mathcal{N}_\xi, \quad U = \frac{1}{\mathcal{N}} \sum_\xi E_\xi \mathcal{N}_\xi. \tag{3.17.2}$$

The number of ways $\mathcal{W}$ in which the ensemble of systems, specified by $\mathcal{N}_\xi$, can be realized (i.e., the number of microstates or thermodynamic probability) can be calculated in exactly the same way as was done above in the case of Boltzmann statistics. As the systems we are considering are distinguishable, one obtains

$$\mathcal{W} = \frac{\mathcal{N}!}{\prod_\xi \mathcal{N}_\xi!}. \tag{3.17.3}$$

Since the number of systems in the ensemble $\mathcal{N}$ and thus all $\mathcal{N}_\xi$ can be arbitrarily high, the actual state of the ensemble is maximizing $\mathcal{W}$ under the constraints of Eq. (3.17.2). Within the method of Lagrange multipliers, one has to maximize the target function

$$\Phi(\mathcal{N}_1, \mathcal{N}_2, \ldots) = \ln \mathcal{W} + \alpha \sum_\xi N_\xi \mathcal{N}_\xi - \beta \sum_\xi E_\xi \mathcal{N}_\xi \tag{3.17.4}$$

(c.f. Eq. (3.6.1) and the following). Using the Stirling formula, Eq. (3.3.1), one obtains

$$\Phi \cong \ln \mathcal{N}! - \sum_\xi \mathcal{N}_\xi \ln \mathcal{N}_\xi + \sum_\xi \mathcal{N}_\xi + \alpha \sum_\xi N_\xi \mathcal{N}_\xi - \beta \sum_\xi E_\xi \mathcal{N}_\xi. \tag{3.17.5}$$

Minimizing with respect to $\mathcal{N}_\xi$, considering $\mathcal{N}$ as a constant, one obtains the equation

$$\frac{\partial \Phi}{\partial \mathcal{N}_\xi} = -\ln \mathcal{N}_\xi + \alpha N_\xi - \beta E_\xi = 0, \tag{3.17.6}$$

which yield

$$\mathcal{N}_\xi = e^{\alpha N_\xi - \beta E_\xi}, \tag{3.17.7}$$

the Gibbs distribution of the so-called grand canonical ensemble.

The total number of systems $\mathcal{N}$ is given by

$$\mathcal{N} = \sum_\xi \mathcal{N}_\xi = \sum_\xi e^{\alpha N_\xi - \beta E_\xi} = \mathcal{Z}, \tag{3.17.8}$$

where $\mathcal{Z}$ is the partition function of the grand canonical ensemble. The coefficients α and β are defined from the conditions for the average number of particles and average energy over the ensemble, Eq. (3.17.2). For the average number of particles, one obtains

$$N = \frac{1}{\mathcal{N}} \sum_\xi N_\xi \mathcal{N}_\xi = \frac{1}{\mathcal{Z}} \sum_\xi N_\xi e^{\alpha N_\xi - \beta E_\xi} = \frac{\partial \ln \mathcal{Z}}{\partial \alpha}. \tag{3.17.9}$$

The ensemble average of the internal energy becomes

$$U = \frac{1}{\mathcal{Z}} \sum_\xi E_\xi e^{\alpha N_\xi - \beta E_\xi} = -\frac{\partial \ln \mathcal{Z}}{\partial \beta}. \tag{3.17.10}$$

In contrast to Eq. (3.6.11), this formula does not contain N explicitly since the average number of particles in a system is encapsulated in $\mathcal{Z}$.

We define the statistical entropy per system as

$$S = \frac{1}{\mathcal{N}} k_B \ln W = \frac{1}{\mathcal{N}} k_B \ln \frac{\mathcal{N}!}{\prod_\xi \mathcal{N}_\xi!}. \tag{3.17.11}$$

Using the Stirling formula, Eq. (3.3.1) without the irrelevant prefactor, one obtains

$$\frac{S}{k_B} = \ln \mathcal{N} - 1 - \frac{1}{\mathcal{N}} \sum_\xi \mathcal{N}_\xi \ln \mathcal{N}_\xi + \frac{1}{\mathcal{N}} \sum_\xi \mathcal{N}_\xi$$

$$= \ln \mathcal{N} - \frac{1}{\mathcal{N}} \sum_\xi \mathcal{N}_\xi \ln \mathcal{N}_\xi. \tag{3.17.12}$$

Inserting here the Gibbs distribution, Eq. (3.17.7), results in

$$\frac{S}{k_B} = \ln \mathcal{N} - \frac{1}{\mathcal{N}} \sum_\xi \mathcal{N}_\xi (\alpha N_\xi - \beta E_\xi) = \ln \mathcal{Z} - \alpha N + \beta U.$$

$$\tag{3.17.13}$$

To clarify the physical meaning of the parameters α and β, one can calculate the differential dS with respect to these parameters and compare the result with the thermodynamic formula for the systems with a variable number of particles and $V = \text{const.}$,

$$dS = \frac{1}{T}dU - \frac{\mu}{T}dN, \tag{3.17.14}$$

where μ is the chemical potential. One obtains

$$\frac{1}{k_B}\frac{\partial S}{\partial \alpha} = \frac{\partial \ln \mathcal{Z}}{\partial \alpha} - N - \alpha\frac{\partial N}{\partial \alpha} + \beta\frac{\partial U}{\partial \alpha} = -\alpha\frac{\partial N}{\partial \alpha} + \beta\frac{\partial U}{\partial \alpha},$$

$$\frac{1}{k_B}\frac{\partial S}{\partial \beta} = \frac{\partial \ln \mathcal{Z}}{\partial \beta} + U - \alpha\frac{\partial N}{\partial \beta} + \beta\frac{\partial U}{\partial \beta} = -\alpha\frac{\partial N}{\partial \beta} + \beta\frac{\partial U}{\partial \beta}. \tag{3.17.15}$$

Combining these formulas and using

$$dN = \frac{\partial N}{\partial \alpha}d\alpha + \frac{\partial N}{\partial \beta}d\beta, \quad dU = \frac{\partial U}{\partial \alpha}d\alpha + \frac{\partial U}{\partial \beta}d\beta, \tag{3.17.16}$$

one obtains

$$dS = k_B\frac{\partial S}{\partial \alpha}d\alpha + k_B\frac{\partial S}{\partial \beta}d\beta = -k_B\alpha dN + k_B\beta dU. \tag{3.17.17}$$

Comparing this formula with Eq. (3.17.14), one identifies

$$\beta = \frac{1}{k_B T}, \quad \alpha = \frac{\mu}{k_B T} = \beta\mu. \tag{3.17.18}$$

Thus, Eq. (3.17.10) can be rewritten as

$$U = -\frac{\partial \ln \mathcal{Z}}{\partial T}\frac{\partial T}{\partial \beta} = k_B T^2\frac{\partial \ln \mathcal{Z}}{\partial T}. \tag{3.17.19}$$

Substituting the values of α and β into Eq. (3.17.13), one obtains

$$-k_B T \ln \mathcal{Z} = U - TS - \mu N. \tag{3.17.20}$$

Using the thermodynamic formulas

$$\mu N = G = U - TS + PV, \tag{3.17.21}$$

one finally obtains

$$-k_B T \ln \mathcal{Z} = -PV = \Omega, \tag{3.17.22}$$

where $\Omega(T, V, \mu)$ is the Ω-potential. Using the main thermodynamic relation for Ω, one can obtain all thermodynamic quantities from $\ln \mathcal{Z}$.

3.18 Statistics of noninteracting indistinguishable particles

The Gibbs distribution looks similar to the Boltzmann distribution, Eq. (3.6.6). However, this is a distribution of systems of particles in the grand canonical ensemble rather than the distribution of particles over their individual energy levels. For systems of noninteracting particles, the individual energy levels i exist, and the distribution of particles over them can be found. For these systems, the state ξ is specified by the set $\{N_i\}$ of the particles' population numbers satisfying

$$N_\xi = \sum_i N_i, \quad E_\xi = \sum_i \varepsilon_i N_i. \qquad (3.18.1)$$

Any state obtained by the redistribution of the particles over the one-particle states i does not count as a new state, only the numbers N_i matter. This implies that within the approach of the grand canonical ensemble, the particles are considered as indistinguishable.

The average of N_i over the grand canonical ensemble is given by

$$\overline{N}_i = \frac{1}{\mathcal{Z}} \sum_\xi N_i \mathcal{N}_\xi = \frac{1}{\mathcal{Z}} \sum_\xi N_i e^{\alpha N_\xi - \beta E_\xi}, \qquad (3.18.2)$$

where $\mathcal{Z}$ is defined by Eq. (3.17.8). As both N_ξ and E_ξ are additive in N_i (see Eq. (3.18.1)), the summand of Eq. (3.17.8) is multiplicative:

$$\mathcal{Z} = \sum_{N_1} e^{(\alpha - \beta \varepsilon_1) N_1} \times \sum_{N_2} e^{(\alpha - \beta \varepsilon_2) N_2} \times \cdots = \prod_j \mathcal{Z}_j. \qquad (3.18.3)$$

As similar factorization occurs in Eq. (3.18.2), all factors except the factor containing N_i cancel each other, leading to

$$\overline{N}_i = \frac{1}{\mathcal{Z}_i} \sum_{N_i} N_i e^{(\alpha - \beta \varepsilon_i) N_i} = \frac{\partial \ln \mathcal{Z}_i}{\partial \alpha}. \qquad (3.18.4)$$

The partition functions for quantum states i have different forms for bosons and fermions. For fermions, N_i, take the values 0 and 1 only, thus

$$Z_i = 1 + e^{\alpha - \beta \varepsilon_i}. \tag{3.18.5}$$

For bosons, N_i, take any values from 0 to ∞, thus

$$Z_i = \sum_{N_i=0}^{\infty} e^{(\alpha - \beta \varepsilon_i)N_i} = \frac{1}{1 - e^{\alpha - \beta \varepsilon_i}}. \tag{3.18.6}$$

Now, $\overline{N}_i$ can be calculated from Eq. (3.18.4). Discarding the bar over N_i, one obtains

$$N_i = \frac{1}{e^{\beta \varepsilon_i - \alpha} \pm 1} = \frac{1}{e^{\beta(\varepsilon_i - \mu)} \pm 1}, \tag{3.18.7}$$

where $(+)$ corresponds to fermions, $(-)$ corresponds to bosons, and α was expressed via μ using Eq. (3.17.18). In the first case, the distribution is called Fermi–Dirac distribution, and in the second case, it is called the Bose–Einstein distribution. Equation (3.18.7) should be compared with the Boltzmann distribution, Eq. (3.6.6). The chemical potential μ is defined by the normalization condition

$$N = \sum_i \frac{1}{e^{\beta(\varepsilon_i - \mu)} \pm 1}. \tag{3.18.8}$$

Replacing summation by integration, this can be rewritten as

$$N = \int_0^{\infty} d\varepsilon \rho(\varepsilon) f(\varepsilon, \mu), \quad f(\varepsilon, \mu) = \frac{1}{e^{\beta(\varepsilon - \mu)} \pm 1}. \tag{3.18.9}$$

The internal energy U is given by

$$U = \sum_i \frac{\varepsilon_i}{e^{\beta(\varepsilon_i - \mu)} \pm 1} \Rightarrow \int_0^{\infty} d\varepsilon \rho(\varepsilon) \varepsilon f(\varepsilon, \mu). \tag{3.18.10}$$

At high temperatures, μ becomes large negative, so that $-\beta\mu \gg 1$. In this case, one can neglect ± 1 in the denominator in comparison to the large exponential, and the Bose–Einstein and

Fermi–Dirac distributions simplify to the Boltzmann distribution. Replacing summation by integration, one obtains

$$N = e^{\beta\mu} \int_0^\infty d\varepsilon \rho(\varepsilon) e^{-\beta\varepsilon} = e^{\beta\mu} Z. \qquad (3.18.11)$$

Using the result for the one-particle partition function for particles in a box, Eq. (3.8.1), yields

$$e^{-\beta\mu} = \frac{Z}{N} = \frac{1}{n}\left(\frac{mk_BT}{2\pi\hbar^2}\right)^{3/2} \gg 1, \qquad (3.18.12)$$

and our calculation is self-consistent. At low temperatures, ± 1 in the denominator of Eq. (3.18.9) cannot be neglected, and the system is called *degenerate*.

The Ω-potential is given by Eq. (3.17.22), which upon substituting Eqs. (3.18.3), (3.18.5), and (3.18.6) becomes

$$\Omega = -k_BT \ln \mathcal{Z} = -k_BT \sum_i \ln \mathcal{Z}_i = \mp k_BT \sum_i \ln(1 \pm e^{\beta(\mu-\varepsilon_i)}),$$

$$(3.18.13)$$

where the upper and lower signs correspond to the Fermi and Bose statistics. Replacing summation by integration, one obtains

$$\Omega = \mp k_BT \int d\varepsilon \rho(\varepsilon) \ln(1 \pm e^{\beta(\mu-\varepsilon)}). \qquad (3.18.14)$$

For the density of states in the form of a power of the energy, $\rho(\varepsilon) = \rho_0 \varepsilon^m$, one can integrate by parts:

$$\Omega = \mp k_BT \frac{\rho_0 \varepsilon^{m+1}}{m+1} \ln(1 \pm e^{\beta(\mu-\varepsilon)}) \Big|_0^\infty - \int d\varepsilon \frac{\rho_0 \varepsilon^{m+1}}{m+1} \frac{1}{e^{\beta(\varepsilon-\mu)} \pm 1},$$

$$(3.18.15)$$

where the first term vanishes and the second term is proportional to the energy given by Eq. (3.18.10):

$$\Omega = -\frac{1}{m+1} U. \qquad (3.18.16)$$

For nonrelativistic particles, $m = 1/2$, see Eq. (3.7.20), so that one obtains the relation

$$PV = \frac{2}{3} U. \qquad (3.18.17)$$

In particular, for the Maxwell–Boltzmann gas, $U = (3/2)\,Nk_BT$, so that the equation of state $PV = Nk_BT$ results. For ultrarelativistic particles, there is a linear relation between the energy and momentum, as for phonons and photons. This implies $m = 2$, see Eq. (3.11.5), keeping in mind $\varepsilon = \hbar\omega$. In this case, one obtains the relation

$$PV = \frac{1}{3}U. \tag{3.18.18}$$

3.19 Bose–Einstein gas

In the macroscopic limit, $N \to \infty$ and $V \to \infty$, so that the concentration of particles $n = N/V$ is constant, the energy levels of the system become so finely quantized that they become quasicontinuous. In this case, summation in Eq. (3.18.8) can be replaced by integration. However, it turns out that in 3D below some temperature T_B, the ideal Bose gas undergoes the so-called Bose condensation. The latter means that a macroscopic number of particles $N_0 \sim N$ falls into the ground state. These particles are called the *Bose condensate*. Setting the energy of the ground state to zero (which can always be done) from Eq. (3.18.9), one obtains

$$N_0 = \frac{1}{e^{-\beta\mu} - 1}. \tag{3.19.1}$$

Resolving this for μ, one obtains

$$\mu = -k_BT \ln\left(1 + \frac{1}{N_0}\right) \cong -\frac{k_BT}{N_0}. \tag{3.19.2}$$

Since N_0 is very large, one has practically $\mu = 0$ below the Bose condensation temperature. Having $\mu = 0$, one can easily calculate the number of particles in the excited states N_{ex} by integration as

$$N_{\text{ex}} = \int_0^\infty d\varepsilon \frac{\rho(\varepsilon)}{e^{\beta\varepsilon} - 1}. \tag{3.19.3}$$

The total number of particles is then

$$N = N_0 + N_{\text{ex}}. \tag{3.19.4}$$

In 3D, using the density of states $\rho(\varepsilon)$ given by Eq. (3.7.20), one obtains

$$
\begin{aligned}
N_{\text{ex}} &= \frac{V}{(2\pi)^2} \left(\frac{2m}{\hbar^2}\right)^{3/2} \int_0^\infty d\varepsilon \frac{\sqrt{\varepsilon}}{e^{\beta\varepsilon} - 1} \\
&= \frac{V}{(2\pi)^2} \left(\frac{2m}{\hbar^2}\right)^{3/2} \frac{1}{\beta^{3/2}} \int_0^\infty dx \frac{\sqrt{x}}{e^x - 1}.
\end{aligned} \tag{3.19.5}
$$

The integral here is a number, a particular case of the general integral

$$
\int_0^\infty dx \frac{x^{s-1}}{e^x - 1} = \Gamma(s)\zeta(s), \tag{3.19.6}
$$

where $\Gamma(s)$ is the gamma function satisfying

$$
\Gamma(n+1) = n\Gamma(n) = n!, \tag{3.19.7}
$$

$$
\Gamma(1/2) = \sqrt{\pi}, \quad \Gamma(3/2) = \sqrt{\pi}/2, \tag{3.19.8}
$$

and $\zeta(s)$ is the Riemann zeta function

$$
\zeta(s) \equiv \sum_{k=1}^\infty k^{-s}, \tag{3.19.9}
$$

having the values

$$
\zeta(1) = \infty, \quad \zeta(3/2) = 2.612, \quad \zeta(5/2) = 1.341, \quad \frac{\zeta(5/2)}{\zeta(3/2)} = 0.5134. \tag{3.19.10}
$$

Thus, Eq. (3.19.5) yields

$$
N_{\text{ex}} = \frac{V}{(2\pi)^2} \left(\frac{2mk_BT}{\hbar^2}\right)^{3/2} \Gamma(3/2)\zeta(3/2), \tag{3.19.11}
$$

increasing with temperature. At $T = T_B$, one has $N_{\text{ex}} = N$, i.e.,

$$
N = \frac{V}{(2\pi)^2} \left(\frac{2mk_BT_B}{\hbar^2}\right)^{3/2} \Gamma(3/2)\zeta(3/2), \tag{3.19.12}
$$

and thus, the condensate disappears, $N_0 = 0$. From the above equation follows

$$
k_BT_B = \frac{\hbar^2}{2m} \left(\frac{(2\pi)^2 n}{\Gamma(3/2)\zeta(3/2)}\right)^{2/3}, \quad n = \frac{N}{V}, \tag{3.19.13}
$$

i.e., $T_B \propto n^{2/3}$. In typical situations, $T_B < 0.1$ K, which is a very low temperature that can be obtained by special methods, such as laser cooling. Now, obviously, Eq. (3.19.11) can be rewritten as

$$\frac{N_{\text{ex}}}{N} = \left(\frac{T}{T_B}\right)^{3/2}, \quad T \leq T_B. \tag{3.19.14}$$

The temperature dependence of the Bose condensate is given by

$$\frac{N_0}{N} = 1 - \frac{N_{\text{ex}}}{N} = 1 - \left(\frac{T}{T_B}\right)^{3/2}, \quad T \leq T_B. \tag{3.19.15}$$

One can see that at $T = 0$, all bosons are in the Bose condensate, while at $T = T_B$, the Bose condensate disappears. Indeed, in the whole temperature range, $0 \leq T \leq T_B$, one has $N_0 \sim N$, and our calculations using $\mu = 0$ in this temperature range are self-consistent. The results above are shown in Fig. 3.10.

Above T_B, there is no Bose condensate, thus $\mu \neq 0$ and is defined by the equation

$$N = \int_0^\infty d\varepsilon \frac{\rho(\varepsilon)}{e^{\beta(\varepsilon - \mu)} - 1}. \tag{3.19.16}$$

This is a nonlinear equation that has no general analytical solution. The result of its numerical solution is shown in Fig. 3.11. At high

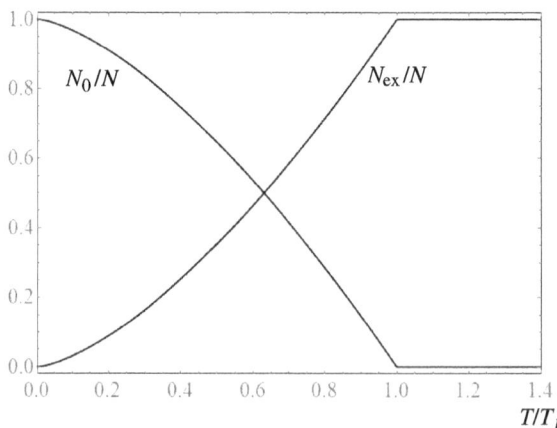

Figure 3.10. Condensate fraction $N_0(T)/N$ and the fraction of excited particles $N_{\text{ex}}(T)/N$ for the ideal Bose–Einstein gas.

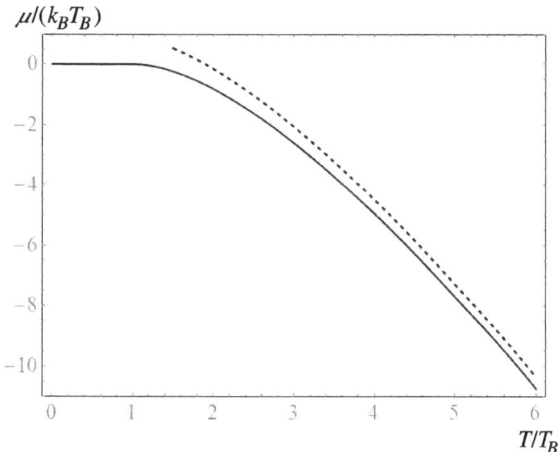

Figure 3.11. Chemical potential $\mu(T)$ of the ideal 3D Bose–Einstein gas ($\mu = 0$ for $T < T_B$) computed numerically by solving Eq. (3.19.16) for $T > T_B$. *Dashed line*: high-temperature asymptote corresponding to the Boltzmann statistics, Eq. (3.18.12).

temperatures, the Bose–Einstein (and also Fermi–Dirac) distribution simplifies to the Boltzmann distribution, and μ can be easily found. Equation (3.18.12) with the help of Eq. (3.19.13) can be rewritten in the form

$$e^{-\beta\mu} = \frac{1}{\zeta(3/2)} \left(\frac{T}{T_B}\right)^{3/2} \gg 1, \qquad (3.19.17)$$

so that the high-temperature limit implies $T \gg T_B$.

Let us now consider the energy of the ideal Bose gas. Since the energy of the condensate is zero, in the thermodynamic limit, one has

$$U = \int_0^\infty d\varepsilon \frac{\varepsilon \rho(\varepsilon)}{e^{\beta(\varepsilon-\mu)} - 1} \qquad (3.19.18)$$

in the whole temperature range. At high temperatures, $T \gg T_B$, the Boltzmann distribution applies, so that U is given by Eq. (3.8.7) and the heat capacity is a constant, $C_V = (3/2)\,Nk_B$. For $T < T_B$, one has $\mu = 0$, thus

$$U = \int_0^\infty d\varepsilon \frac{\varepsilon \rho(\varepsilon)}{e^{\beta\varepsilon} - 1}, \qquad (3.19.19)$$

which in three dimensions becomes

$$
U = \frac{V}{(2\pi)^2} \left(\frac{2m}{\hbar^2}\right)^{3/2} \int_0^\infty d\varepsilon \frac{\varepsilon^{3/2}}{e^{\beta\varepsilon} - 1}
$$

$$
= \frac{V}{(2\pi)^2} \left(\frac{2m}{\hbar^2}\right)^{3/2} \Gamma(5/2)\zeta(5/2) \, (k_B T)^{5/2} . \tag{3.19.20}
$$

With the help of Eq. (3.19.12), this can be expressed in the form

$$
U = N k_B T \left(\frac{T}{T_B}\right)^{3/2} \frac{\Gamma(5/2)\zeta(5/2)}{\Gamma(3/2)\zeta(3/2)}
$$

$$
= N k_B T \left(\frac{T}{T_B}\right)^{3/2} \frac{3}{2} \frac{\zeta(5/2)}{\zeta(3/2)}, \quad T \le T_B. \tag{3.19.21}
$$

The corresponding heat capacity is

$$
C_V = \left(\frac{\partial U}{\partial T}\right)_V = N k_B \left(\frac{T}{T_B}\right)^{3/2} \frac{15}{4} \frac{\zeta(5/2)}{\zeta(3/2)}. \tag{3.19.22}
$$

To find the heat capacity at $T > T_B$, one first has to find $\mu(T)$ numerically from Eq. (3.19.16), substitute it into Eq. (3.19.18) to find the energy, and then differentiate the result with respect to T. The result for the heat capacity of the 3D Bose gas in the whole range of the temperature is shown in Fig. 3.12. In the high-temperature limit, quantum effects disappear and the classical asymptotic value $C_V/N = 3/2 k_B$ is reached.

To calculate the pressure of the ideal Bose gas, one can start with the entropy, which below T_B is given by

$$
S = \int_0^T dT' \frac{C_V(T')}{T'} = \frac{2}{3} C_V = N k_B \left(\frac{T}{T_B}\right)^{3/2} \frac{5}{2} \frac{\zeta(5/2)}{\zeta(3/2)}. \tag{3.19.23}
$$

Differentiating the free energy

$$
F = U - TS = -N k_B T \left(\frac{T}{T_B}\right)^{3/2} \frac{\zeta(5/2)}{\zeta(3/2)}, \tag{3.19.24}
$$

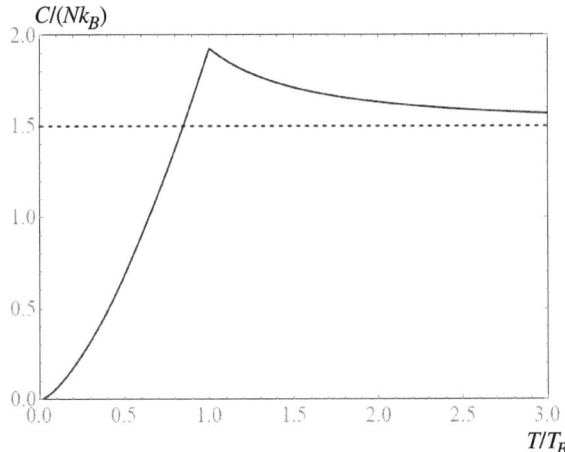

Figure 3.12. Heat capacity $C_V(T)$ of the ideal 3D Bose–Einstein gas.

one obtains the pressure

$$P = -\left(\frac{\partial F}{\partial V}\right)_T = -\left(\frac{\partial F}{\partial T_B}\right)_T \left(\frac{\partial T_B}{\partial V}\right)_T$$

$$= -\left(-\frac{3}{2}\frac{F}{T_B}\right)\left(-\frac{2}{3}\frac{T_B}{V}\right) = -\frac{F}{V}, \qquad (3.19.25)$$

i.e., the equation of state

$$PV = Nk_BT\left(\frac{T}{T_B}\right)^{3/2}\frac{\zeta(5/2)}{\zeta(3/2)}, \quad T \le T_B, \qquad (3.19.26)$$

compared to $PV = Nk_BT$ at high temperatures. In fact, this result can be obtained immediately from the pressure–energy relation, Eq. (3.18.17).

One can see that the pressure of the ideal Bose gas with the condensate contains an additional factor $(T/T_B)^{3/2} < 1$, the fraction of particles in the excited states. This is because the particles in the condensate are not thermally agitated, and thus, they do not contribute to the pressure. Using Eq. (3.19.14), one can rewrite the equation of state of the 3D Bose gas at $T \le T_B$ as

$$PV = N_{\text{ex}}(T)k_BT\frac{\zeta(5/2)}{\zeta(3/2)}. \qquad (3.19.27)$$

Other thermodynamic functions above can also be rewritten in this way. Combining this with Eq. (3.19.11), one can see that the volume cancels out in the equation of state and the pressure is a function of the temperature alone:

$$P = P_B(T) \equiv \frac{\Gamma(3/2)\zeta(5/2)}{(2\pi)^2} \left(\frac{2mk_BT}{\hbar^2} \right)^{3/2} k_BT. \qquad (3.19.28)$$

This is the same behavior as that of the saturated vapor. We can see that one cannot define the heat capacity at constant pressure C_P as one cannot change T keeping $P = \text{const.}$ Also, one cannot define the meaningful thermal expansion coefficient and compressibility for the Bose gas below T_B. If the external pressure exceeds the pressure of the gas above, the gas will be compressed to zero volume. In the opposite case, the gas will expand until the condensate disappears, and then the volume stabilizes at some finite value.

3.20   Fermi–Dirac gas

Properties of the Fermi gas (plus sign in Eq. (3.18.9)) are different from those of the Bose gas because the exclusion principle prevents multioccupancy of quantum states. As a result, no condensation at the ground state occurs at low temperatures. For a macroscopic system, the chemical potential can be found from the equation

$$N = \int_0^\infty d\varepsilon \rho(\varepsilon) f(\varepsilon) = \int_0^\infty d\varepsilon \frac{\rho(\varepsilon)}{e^{\beta(\varepsilon-\mu)} + 1} \qquad (3.20.1)$$

at all temperatures. This is a nonlinear equation for μ that in general can be solved only numerically. The result of the numerical solution for $\mu(T)$ is shown in Fig. 3.13. The Fermi–Dirac distribution function for different temperatures is shown in Fig. 3.14.

In the limit, $T \to 0$, fermions fill a certain number of low-lying energy levels to minimize the total energy while obeying the exclusion principle. As we will see, the chemical potential of fermions is positive at low temperatures, $\mu > 0$. For $T \to 0$ (i.e., $\beta \to \infty$), one has $e^{\beta(\varepsilon-\mu)} \to 0$ if $\varepsilon < \mu$ and $e^{\beta(\varepsilon-\mu)} \to \infty$ if $\varepsilon > \mu$. Thus, Eq. (3.20.1)

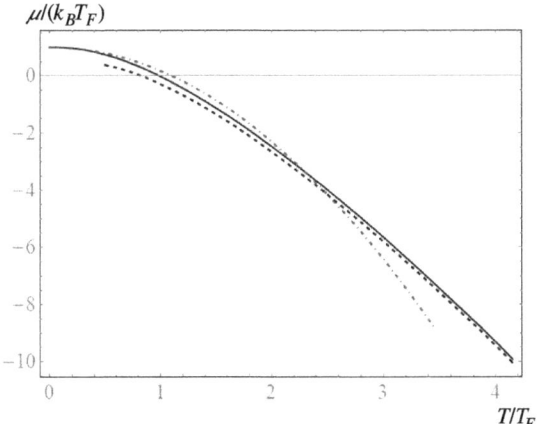

Figure 3.13. Chemical potential $\mu(T)$ of the ideal Fermi–Dirac gas. *Dashed line*: high-temperature asymptote corresponding to the Boltzmann statistics, Eq. (3.18.12). *Dashed-dotted line*: low temperature asymptote, Eq. (3.20.16).

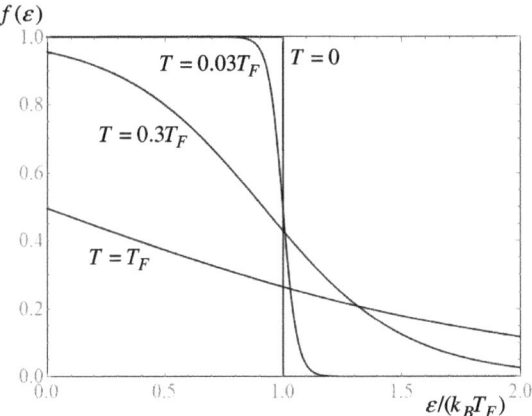

Figure 3.14. Fermi–Dirac distribution function at different temperatures.

becomes

$$f(\varepsilon) = \begin{cases} 1, & \varepsilon < \mu, \\ 0, & \varepsilon > \mu. \end{cases} \tag{3.20.2}$$

The zero-temperature value of μ that we call the Fermi energy ε_F is defined by the equation

$$N = \int_0^{\varepsilon_F} d\varepsilon \rho(\varepsilon). \tag{3.20.3}$$

Fermions are mainly electrons having spin 1/2 and correspondingly degeneracy 2 because of the two states of the spin. In three dimensions, using Eq. (3.7.20) with an additional factor 2 for the degeneracy, one obtains

$$N = \frac{2V}{(2\pi)^2} \left(\frac{2m}{\hbar^2}\right)^{3/2} \int_0^{\varepsilon_F} d\varepsilon \sqrt{\varepsilon} = \frac{2V}{(2\pi)^2} \left(\frac{2m}{\hbar^2}\right)^{3/2} \frac{2}{3}\varepsilon_F^{3/2}. \quad (3.20.4)$$

From here, one obtains

$$\varepsilon_F = \frac{\hbar^2}{2m} \left(3\pi^2 n\right)^{2/3}. \quad (3.20.5)$$

It is also convenient to introduce the Fermi temperature as

$$k_B T_F = \varepsilon_F. \quad (3.20.6)$$

One can see that T_F has the same structure as T_B defined by Eq. (3.19.13). In typical metals, $T_F \sim 10^5$ K, so that at room temperatures, $T \ll T_F$ and the electron gas is degenerate. It is convenient to express the density of states in three dimensions, Eq. (3.7.20), in terms of ε_F:

$$\rho(\varepsilon) = \frac{3}{2} N \frac{\sqrt{\varepsilon}}{\varepsilon_F^{3/2}}. \quad (3.20.7)$$

Let us now calculate the internal energy U at $T = 0$:

$$U = \int_0^{\mu_0} d\varepsilon \rho(\varepsilon)\varepsilon = \frac{3}{2}\frac{N}{\varepsilon_F^{3/2}} \int_0^{\varepsilon_F} d\varepsilon\, \varepsilon^{3/2} = \frac{3}{2}\frac{N}{\varepsilon_F^{3/2}} \frac{2}{5}\varepsilon_F^{5/2} = \frac{3}{5} N\varepsilon_F. \quad (3.20.8)$$

One cannot calculate the heat capacity C_V from this formula as it requires taking into account small temperature-dependent corrections in U. This will be done later. One can calculate the pressure at low temperatures since in this region, the entropy S should be small and $F = U - TS \cong U$. One obtains

$$P = -\left(\frac{\partial F}{\partial V}\right)_{T=0} \cong -\left(\frac{\partial U}{\partial V}\right)_{T=0} = -\frac{3}{5}N\frac{\partial \varepsilon_F}{\partial V} = -\frac{3}{5}N\left(-\frac{2}{3}\frac{\varepsilon_F}{V}\right)$$

$$= \frac{2}{5}n\varepsilon_F = \frac{\hbar^2}{2m}\frac{2}{5}\left(3\pi^2\right)^{2/3} n^{5/3}. \quad (3.20.9)$$

To calculate the heat capacity at low temperatures, one has to find temperature-dependent corrections to Eq. (3.20.8). We will need the integral of a general type:

$$M_\eta = \int_0^\infty d\varepsilon \, \varepsilon^\eta f(\varepsilon) = \int_0^\infty d\varepsilon \frac{\varepsilon^\eta}{e^{(\varepsilon-\mu)/(k_BT)} + 1}, \qquad (3.20.10)$$

which enters Eq. (3.20.1) for N and the similar equation for U. With the use of Eq. (3.20.7), one can write

$$N = \frac{3}{2} \frac{N}{\varepsilon_F^{3/2}} M_{1/2}, \qquad (3.20.11)$$

$$U = \frac{3}{2} \frac{N}{\varepsilon_F^{3/2}} M_{3/2}. \qquad (3.20.12)$$

It will be shown at the end of this section that for $k_BT \ll \mu$, the expansion of M_η up to quadratic terms has the form

$$M_\eta = \frac{\mu^{\eta+1}}{\eta+1} \left[1 + \frac{\pi^2 \eta(\eta+1)}{6} \left(\frac{k_BT}{\mu} \right)^2 \right]. \qquad (3.20.13)$$

Now, Eq. (3.20.11) takes the form

$$\varepsilon_F^{3/2} = \mu^{3/2} \left[1 + \frac{\pi^2}{8} \left(\frac{k_BT}{\mu} \right)^2 \right], \qquad (3.20.14)$$

which defines $\mu(T)$ up to the terms of order T^2:

$$\mu = \varepsilon_F \left[1 + \frac{\pi^2}{8} \left(\frac{k_BT}{\mu} \right)^2 \right]^{-2/3} \cong \varepsilon_F \left[1 - \frac{\pi^2}{12} \left(\frac{k_BT}{\mu} \right)^2 \right]$$

$$\cong \varepsilon_F \left[1 - \frac{\pi^2}{12} \left(\frac{k_BT}{\varepsilon_F} \right)^2 \right] \qquad (3.20.15)$$

or, with the help of Eq. (3.20.6),

$$\mu = \varepsilon_F \left[1 - \frac{\pi^2}{12} \left(\frac{T}{T_F} \right)^2 \right]. \qquad (3.20.16)$$

It is not surprising that the chemical potential decreases with temperature because at high temperatures, it takes large negative values, see Eq. (3.18.12). Equation (3.20.12) becomes

$$U = \frac{3}{2}\frac{N}{\varepsilon_F^{3/2}}\frac{\mu^{5/2}}{(5/2)}\left[1 + \frac{5\pi^2}{8}\left(\frac{k_B T}{\mu}\right)^2\right]$$

$$\cong \frac{3}{5}N\frac{\mu^{5/2}}{\varepsilon_F^{3/2}}\left[1 + \frac{5\pi^2}{8}\left(\frac{k_B T}{\varepsilon_F}\right)^2\right]. \qquad (3.20.17)$$

Using Eq. (3.20.16), one obtains

$$U = \frac{3}{5}N\varepsilon_F\left[1 - \frac{\pi^2}{12}\left(\frac{T}{T_F}\right)^2\right]^{5/2}\left[1 + \frac{5\pi^2}{8}\left(\frac{T}{T_F}\right)^2\right]$$

$$\cong \frac{3}{5}N\varepsilon_F\left[1 - \frac{5\pi^2}{24}\left(\frac{T}{T_F}\right)^2\right]\left[1 + \frac{5\pi^2}{8}\left(\frac{T}{T_F}\right)^2\right], \qquad (3.20.18)$$

which yields

$$U = \frac{3}{5}N\varepsilon_F\left[1 + \frac{5\pi^2}{12}\left(\frac{T}{T_F}\right)^2\right]. \qquad (3.20.19)$$

At $T = 0$, this formula reduces to Eq. (3.20.8). Now, one obtains

$$C_V = \left(\frac{\partial U}{\partial T}\right)_V = Nk_B\frac{\pi^2}{2}\frac{T}{T_F}, \qquad (3.20.20)$$

which is small at $T \ll T_F$. The numerically found heat capacity in the whole temperature range is shown in Fig. 3.15. In the high-temperature limit, it tends to the classical result $C_V/N = 3/2k_B$.

Let us now derive Eq. (3.20.13). Integrating Eq. (3.20.10) by parts, one obtains

$$M_\eta = \frac{\varepsilon^{\eta+1}}{\eta+1}f(\varepsilon)\Big|_0^\infty - \int_0^\infty d\varepsilon\,\frac{\varepsilon^{\eta+1}}{\eta+1}\frac{\partial f(\varepsilon)}{\partial\varepsilon}. \qquad (3.20.21)$$

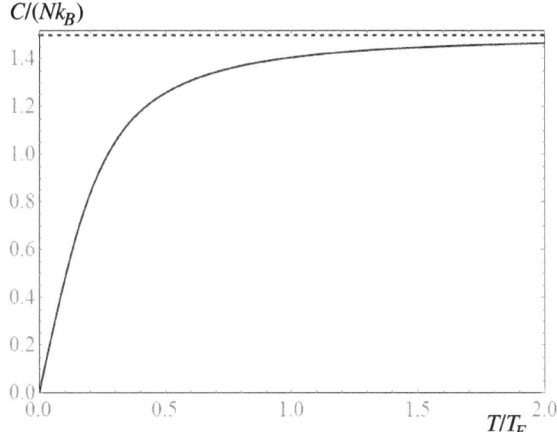

Figure 3.15. Heat capacity $C(T)$ of the ideal Fermi–Dirac gas.

The first term of this formula is zero. At low temperatures, $f(\varepsilon)$ is close to a step function, fast changing from 1 to 0 in the vicinity of $\varepsilon = \mu$. Thus,

$$\frac{\partial f(\varepsilon)}{\partial \varepsilon} = \frac{\partial}{\partial \varepsilon} \frac{1}{e^{\beta(\varepsilon - \mu)} + 1} = -\frac{\beta e^{\beta(\varepsilon - \mu)}}{\left[e^{\beta(\varepsilon - \mu)} + 1\right]^2} = -\frac{\beta}{4 \cosh^2 \left[\beta \left(\varepsilon - \mu\right)/2\right]}$$

(3.20.22)

has a sharp negative peak at $\varepsilon = \mu$. On the contrary, $\varepsilon^{\eta+1}$ is a slow function of ε that can be expanded in the Taylor series in the vicinity of $\varepsilon = \mu$. Up to the second order, one has

$$\varepsilon^{\eta+1} = \mu^{\eta+1} + \left.\frac{\partial \varepsilon^{\eta+1}}{\partial \varepsilon}\right|_{\varepsilon = \mu} (\varepsilon - \mu) + \frac{1}{2} \left.\frac{\partial^2 \varepsilon^{\eta+1}}{\partial \varepsilon^2}\right|_{\varepsilon = \mu} (\varepsilon - \mu)^2$$

$$= \mu^{\eta+1} + (\eta + 1) \mu^{\eta} (\varepsilon - \mu) + \frac{1}{2} \eta (\eta + 1) \mu^{\eta-1} (\varepsilon - \mu)^2.$$

(3.20.23)

Introducing $x \equiv \beta \left(\varepsilon - \mu\right)/2$ and formally extending integration in Eq. (3.20.21) from $-\infty$ to ∞, one obtains

$$M_\eta = \frac{\mu^{\eta+1}}{\eta + 1} \int_{-\infty}^{\infty} dx \left[\frac{1}{2} + \frac{\eta + 1}{\beta \mu} x + \frac{\eta (\eta + 1)}{\beta^2 \mu^2} x^2\right] \frac{1}{\cosh^2 (x)}.$$

(3.20.24)

The contribution of the linear x term vanishes by symmetry. Using the integrals

$$\int_{-\infty}^{\infty} dx \frac{1}{\cosh^2(x)} = 2, \quad \int_{-\infty}^{\infty} dx \frac{x^2}{\cosh^2(x)} = \frac{\pi^2}{6}, \qquad (3.20.25)$$

one arrives at Eq. (3.20.13).

3.21   Electromagnetic radiation (photon gas)

The heat propagating away from hot objects is electromagnetic radiation. It was experimentally established that in a cavity within a body, there is electromagnetic radiation at equilibrium with the body, whose intensity and average frequency increase with the temperature of the body. If a small hole is bored in the body connecting the cavity with the outside world, a stream of electromagnetic radiation goes out and can be analyzed. The analysis of its intensity and frequency distribution led to the introduction of the concept of quantization in physics by Planck in 1900.

The properties of the equilibrium electromagnetic radiation are similar to those of elastic vibrations in solids. Quanta of the latter are called *phonons*, while those of the former are called *photons*. So, instead of electromagnetic radiation, one can speak of the photon gas. It is clear that the properties of the photon gas do not depend on the shape of the cavity (that was also proven experimentally). Thus, one can consider a cavity in the form of a box with the sizes L_x, L_y, and L_z, applying boundary conditions at the borders of the cavity. The most natural boundary conditions plausible in the case of metals require that the electric field in the electromagnetic wave be zero at the borders, otherwise a strong current will be induced. This boundary condition leads to the normal modes in the form of standing plane waves with the wave vectors $\mathbf{k} = \{k_x, k_y, k_z\}$ given by

$$k_\alpha = \frac{\pi}{L_\alpha} \nu_\alpha, \quad \nu_\alpha = 1, 2, 3, \ldots, \quad \alpha = x, y, z. \qquad (3.21.1)$$

This formula is similar to those for a quantum particle in a box, Eq. (3.7.12), and for a clamped elastic body, Eq. (3.11.2). An important difference with the elastic problem is that the values of the wave vectors are unlimited.

The frequency density of the modes can be obtained in the same way as was done for the elastic waves. There are only two transverse polarizations of the electromagnetic waves instead of three polarizations of the elastic waves. Taking this into account and using $\omega = ck$, where c is the speed of light, one obtains

$$\rho(\omega) = \frac{V}{\pi^2 c^3} \omega^2, \tag{3.21.2}$$

where $V = L_x L_y L_z$ is the volume of the cavity.

In fact, the properties of the electromagnetic radiation (also called *black-body radiation*) were experimentally investigated before those of the elastic system. First, the problem was considered classically, and it was supposed that there is the energy equal to $k_B T$ in each mode. Multiplying the formula above by $k_B T$ and dividing it by V, one obtains the spectral energy density

$$\rho_E(\omega) = \frac{k_B T}{\pi^2 c^3} \omega^2, \tag{3.21.3}$$

the Rayleigh–Jeans formula. The integral of Eq. (3.21.3) over the frequencies, with the upper bound equal to infinity, is diverging, which is termed the *ultraviolet catastrophe*. Experiments, however, showed that this formula is valid only at low frequencies. The experimentally found spectral energy density had a maximum at some temperature-dependent frequency and decreased exponentially at high frequencies. Trying to understand this, Planck proposed the concept of quantization of the energy of a harmonic oscillator, in this case, the energy of an electromagnetic wave. Within classical statistical physics, the energy of the oscillator obeys the Boltzmann statistics with the continuous values of the oscillator's energy. The partition function of a single oscillator up to a constant factor is given by

$$Z = \int_0^\infty d\varepsilon e^{-\beta \varepsilon} = \frac{1}{\beta}. \tag{3.21.4}$$

Then, the oscillator's thermal energy becomes

$$E = -\frac{\partial \ln Z}{\partial \beta} = \frac{1}{\beta} = k_B T, \tag{3.21.5}$$

which leads to the Rayleigh–Jeans formula, Eq. (3.21.3). Planck tried quantized values of the oscillator's energies, $\varepsilon = \nu \hbar \omega$ with

$\nu = 0, 1, 2, \ldots$, ω being the oscillator's frequency, and $\hbar$ being a new physical constant that was then called Planck's constant. In this case, the partition function is a geometrical progression:

$$Z = \sum_{\nu=0}^{\infty} (e^{-\beta\hbar\omega})^{\nu} = \frac{1}{1 - e^{-\beta\hbar\omega}}, \qquad (3.21.6)$$

and the oscillator's thermal energy becomes

$$E = -\frac{\partial \ln Z}{\partial \beta} = \frac{\hbar\omega}{e^{\beta\hbar\omega} - 1}, \qquad (3.21.7)$$

whereas the average number of quanta $n(\omega) = \langle \nu \rangle$ becomes

$$n(\omega) = \frac{E}{\hbar\omega} = \frac{1}{e^{\beta\hbar\omega} - 1}. \qquad (3.21.8)$$

At high temperatures, $\beta\hbar\omega \ll 1$, the classical result $E = k_B T$ is recovered. In this case, $e^{-\beta\hbar\omega}$ is close to one, and summation can be replaced by integration. In general, one obtains Planck's formula

$$\rho_E(\omega) = \frac{\hbar}{\pi^2 c^3} \frac{\omega^3}{e^{\beta\hbar\omega} - 1}, \qquad (3.21.9)$$

which perfectly describes the experiment. In particular, the maximum of the spectral energy density satisfies the transcendental equation

$$3(1 - e^{-\beta\hbar\omega}) = \beta\hbar\omega, \qquad (3.21.10)$$

thus $\omega_{\max} \propto k_B T/\hbar$. In this way, Planck's constant had been introduced in physics, and it could be measured as $\hbar = 1.055 \times 10^{-34}\,\text{J} \cdot \text{s}$. Planck had empirically found quantum energy levels of the harmonic oscillator before quantum mechanics was created. From Planck's finding, it follows that the energy of the electromagnetic field is split into discrete portions, quanta, with the energy $\hbar\omega$. These quanta of electromagnetic energy are called *photons*. Photons and phonons are examples of *quasiparticles* — excitations having some properties of particles.

The derivation above repeats that for the quantum harmonic oscillator starting from Eq. (3.9.3). The only difference is that the zero-point energy is absent in Planck's derivation. However, as this energy cannot be released, it cannot be measured and has to be discarded in interpreting the energy of the black-body radiation. Integrating the spectral energy density over the frequencies, one obtains the energy of the photon gas:

$$U = u(T)V, \quad u(T) = \int_0^\infty d\omega \rho_E(\omega) = \frac{\pi^2}{15c^3\hbar^3}(k_B T)^4, \quad (3.21.11)$$

which is similar to Eq. (3.11.19) for the phonon gas in the elastic body. The heat capacity is given by

$$C_V = \left(\frac{\partial U}{\partial T}\right)_V = \frac{4\pi^2 V k_B}{15c^3\hbar^3}(k_B T)^3. \quad (3.21.12)$$

Let us calculate the pressure of the photon gas. The entropy can be obtained by the integration:

$$S = \int_0^T dT' \frac{C_V(T')}{T'} = \frac{1}{3}C_V = \frac{4\pi^2 V k_B}{45c^3\hbar^3}(k_B T)^3. \quad (3.21.13)$$

Then, the free energy becomes

$$F = U - TS = U - \frac{4}{3}U = -\frac{1}{3}U = -\frac{1}{3}u(T)V. \quad (3.21.14)$$

Now, the equation of state is given by

$$P = -\left(\frac{\partial F}{\partial V}\right)_T = \frac{1}{3}u(T), \quad (3.21.15)$$

in accordance with Eq. (3.18.18) for ultrarelativistic particles. One can see that the photon gas behaves as saturated vapor — it can be compressed without any pressure increase because of condensation. Indeed, one can calculate the number of photons in the photon gas,

$$N_{ph} = \int_0^\infty d\omega \rho(\omega) n(\omega), \quad (3.21.16)$$

and it turns out to be proportional to the volume and depending on the temperature. Thus, the number of photons is not conserved, unlike that of ordinary particles. The pressure of the photons is small, 2.4×10^{-6} N/m^2 at $T = 300$ K, much smaller than the atmospheric pressure 10^5 N/m^2, still it can be demonstrated in a high-school experiment.

The electromagnetic radiation at equilibrium can be considered in a different way using the statistics of indistinguishable particles based on the grand canonical ensemble. In the description of harmonic oscillators, phonons, and photons above, we considered equidistant quantum energy levels of harmonic oscillators or normal modes that behave like harmonic oscillators, and we applied the Boltzmann statistics to find the number of the oscillators in each of their quantum states. The oscillators or normal modes are obviously distinguishable, but this does not matter. Now, we consider quantum states of normal modes, asking how many excitations (quasiparticles, such as phonons and photons) are in each mode, like how many particles are in a box. Obviously, the number of excitations in each mode is unlimited and the excitations are indistinguishable by definition. One cannot speak about Excitation 1 and Excitation 2, redistributing which between different modes, one could obtain new microstates. Thus, one can apply the method of the grand canonical ensemble and use the Bose–Einstein distribution for the number of excitations in each state. The total number of excitations N is obviously not conserved. Thus, in the grand canonical ensemble, there is no constraint on the number of particles and thus no Lagrange parameter α and no chemical potential μ. As a result, one can use the Bose distribution function with $\mu = 0$:

$$n(\omega) = \frac{1}{e^{\beta \hbar \omega} - 1} \tag{3.21.17}$$

for the number of excitations in the mode (state) with the frequency ω. This is Eq. (3.9.19) obtained for the harmonic oscillator by summing up the geometrical progression and Eq. (3.21.8). Multiplying this by the density of modes (now states) $\rho(\omega)$ of Eq. (3.21.2) for photons and by the energy of the mode (now state) $\hbar \omega$, one arrives at Planck's formula, Eq. (3.21.9).

It should be stressed that the quasiparticle picture of the excited states of the modes is only possible if the excitation spectrum of the mode is equidistant. Then the energy of the mode is proportional to the number of excitations, and one can use the energy constraint in the form of Eq. (3.1.3) or Eq. (3.18.1). If the energy levels are not equidistant, as is the case for the anharmonic oscillator, rotator, or even a particle in a box, one cannot use these formulas for the energy within the quasiparticle description. Still, the approach based on the Boltzmann statistics is working in these cases.

3.22 Problems

3.22.1 Thermodynamics of the Fermi gas

Calculate isothermal compressibility κ_T, thermal expansivity β, and C_P for the ideal Fermi gas at low temperatures.

Solution: In the compressibility, the thermal effect is negligible, and one can use the zero-temperature formula

$$P = \frac{2}{5} n \varepsilon_F = \frac{\hbar^2}{2m} \frac{2}{5} (3\pi^2)^{2/3} n^{5/3} = \frac{\hbar^2}{2m} \frac{2}{5} (3\pi^2)^{2/3} \left(\frac{N}{V}\right)^{5/3}. \tag{3.22.1}$$

Since $V \propto P^{-3/5}$, the compressibility becomes

$$\varkappa = -\frac{1}{V} \frac{\partial V}{\partial P} = \frac{3}{5P} = \frac{3}{2n\varepsilon_F} \propto n^{-5/3}.$$

The more the gas is compressed, the smaller is the compressibility.
Calculation of the thermal expansivity

$$\beta = \frac{1}{V} \left(\frac{\partial V}{\partial T}\right)_P$$

requires the knowledge of the equation of state with thermal corrections. The equation of state is defined by

$$P = -\left(\frac{\partial F}{\partial V}\right)_T, \quad F = U - TS.$$

For the energy, one uses the formula

$$U = \frac{3}{5} N \varepsilon_F \left[1 + \frac{5\pi^2}{12} \left(\frac{T}{T_F}\right)^2\right]. \tag{3.22.2}$$

The entropy can be found as

$$S = \int_0^T dT' \frac{C_V}{T}.$$

Using

$$C_V = \left(\frac{\partial U}{\partial T} \right)_V = N k_B \frac{\pi^2}{2} \frac{T}{T_F}, \tag{3.22.3}$$

one obtains

$$S = C_V = N k_B \frac{\pi^2}{2} \frac{T}{T_F}.$$

Thus,

$$F = \frac{3}{5} N \varepsilon_F + \frac{\pi^2}{4} N k_B \frac{T^2}{T_F} - N k_B \frac{\pi^2}{2} \frac{T^2}{T_F} = \frac{3}{5} N \varepsilon_F - \frac{\pi^2}{4} N k_B \frac{T^2}{T_F}.$$

Now, using $\varepsilon_F \propto T_F \propto V^{-2/3}$, one obtains

$$P = - \left(\frac{\partial F}{\partial V} \right)_T = -\frac{3}{5} N \left(-\frac{2}{3} \frac{\varepsilon_F}{V} \right) + \frac{\pi^2}{4} N k_B \frac{2}{3V} \frac{T^2}{T_F}$$

$$= \frac{2}{5} n \varepsilon_F + \frac{\pi^2}{6} n \frac{(k_B T)^2}{\varepsilon_F}$$

or

$$P = \frac{2}{5} n \varepsilon_F \left[1 + \frac{5\pi^2}{12} \left(\frac{T}{T_F} \right)^2 \right],$$

which is the equation of state with the thermal corrections. That is, the pressure increases with the temperature, as it should be. Again, similar to the Bose gas,

$$U = \frac{3}{2} PV.$$

Now, to calculate the thermal expansivity, one can relate dT and dV using the equation of state above. The differential of P is given by

$$dP = \frac{\partial}{\partial V}\left(\frac{2}{5}n\varepsilon_F + \frac{\pi^2}{6}n\frac{(k_BT)^2}{\varepsilon_F}\right)dV + \frac{\partial}{\partial T}\left(\frac{\pi^2}{6}n\frac{(k_BT)^2}{\varepsilon_F}\right)dT,$$

where in the volume term, the temperature-correction term can be neglected. As $\varepsilon_F \propto V^{-2/3}$, one has

$$dP = \frac{2}{5}\left(-\frac{5}{3}\frac{n\varepsilon_F}{V}\right)dV + \frac{\pi^2}{3}n\frac{k_B^2 T}{\varepsilon_F}dT.$$

Using $dP = 0$, one obtains

$$\beta = \frac{1}{V}\left(\frac{\partial V}{\partial T}\right)_P = \frac{\pi^2}{2}\frac{k_B^2 T}{\varepsilon_F^2} = \frac{\pi^2}{2}\frac{T}{T_F^2},$$

which is small at low temperatures.

To calculate C_P, we use

$$C_P - C_V = T\left(\frac{\partial P}{\partial T}\right)_V\left(\frac{\partial V}{\partial T}\right)_P = VT\frac{\beta^2}{\varkappa_T}. \qquad (3.22.4)$$

Using the above results, one obtains

$$C_P - C_V = VT\frac{\left(\frac{\pi^2}{2}\frac{T}{T_F^2}\right)^2}{\frac{3}{2n\varepsilon_F}} = \frac{3\pi^4}{2}nVTk_BT_F\frac{T^2}{T_F^4} = \frac{3\pi^4}{2}Nk_B\left(\frac{T}{T_F}\right)^3,$$

a very small difference.

3.22.2 The isotherm of the ideal 3D Bose gas

Find the V–P relation at $T = $ const. for the ideal Bose gas in 3D.

Solution: The volume V of the gas will be finite for $P < P_B(T)$, where

$$P_B(T) \equiv \frac{\Gamma(3/2)\zeta(5/2)}{(2\pi)^2}\left(\frac{2mk_BT}{\hbar^2}\right)^{3/2}k_BT. \qquad (3.22.5)$$

In this region, there is no Bose condensate and $\mu < 0$. In the limit of small pressure, the equation of state of the ideal Maxwell–Boltzmann gas

$$PV = Nk_BT$$

should be recovered. At $P = P_B(T)$, Bose condensation begins: $\mu = 0$. Here, the isotherm goes vertically down to zero volume. Changing the volume does not affect the pressure that depends only on the temperature. As the volume decreases, more and more particles fall into the condensate.

Let us set up the model equations for $P < P_B(T)$. First, there is the normalization condition

$$N = \int_0^\infty d\varepsilon \rho(\varepsilon) f(\varepsilon, \mu) = \frac{V}{(2\pi)^2} \left(\frac{2m}{\hbar^2}\right)^{3/2} \int_0^\infty d\varepsilon \frac{\sqrt{\varepsilon}}{e^{\beta(\varepsilon-\mu)} - 1}.$$

There is the full number of particles N on the left as there is no condensate. It is convenient to use the reduced volume

$$\tilde{v} \equiv \frac{V}{V_B}, \quad V_B \equiv \frac{Nk_BT}{P_B(T)}, \tag{3.22.6}$$

where V_B is the volume of the Maxwell–Boltzmann gas at the pressure $P = P_B$. With this notation, the normalization condition becomes

$$1 = \frac{\tilde{v}\beta^{3/2}}{\Gamma(3/2)\zeta(5/2)} \int_0^\infty d\varepsilon \frac{\sqrt{\varepsilon}}{e^{\beta(\varepsilon-\mu)} - 1}. \tag{3.22.7}$$

We need one more equation that includes the pressure. Using the relation

$$U = \frac{3}{2}PV \tag{3.22.8}$$

and the expression for the energy,

$$U = \int_0^\infty d\varepsilon \rho(\varepsilon)\varepsilon f(\varepsilon, \mu),$$

for the reduced pressure,

$$\tilde{p} = \frac{P}{P_B},$$

one obtains

$$\tilde{p} = \frac{2U}{3VP_B} = \frac{\beta^{5/2}}{\Gamma(5/2)\zeta(5/2)} \int_0^\infty d\varepsilon \frac{\varepsilon^{3/2}}{e^{\beta(\varepsilon-\mu)} - 1}, \tag{3.22.9}$$

where we also used $(3/2)\,\Gamma(3/2) = \Gamma(5/2)$. From this equation, one can find μ numerically for a given $\tilde{p}$ and then substitute this result into Eq. (3.22.7) to compute $\tilde{v}\,(\tilde{p})$, which is the isotherm of the Bose gas in the reduced form. Setting $\mu = 0$ in the equation above yields $\tilde{p} = 1$, as it should be in the presence of the condensate. In Eq. (3.22.7), $\mu = 0$ corresponds exactly to the onset of the Bose condensation, as in the condensed phase, there is N_{ex}/N on the left side of this equation. Thus, the corresponding value

$$\tilde{v} = \frac{\zeta(5/2)}{\zeta(3/2)} = 0.5134$$

defines the volume of the Bose gas corresponding to the onset of the Bose condensation. Further compression leads to the decrease of the volume to zero at the constant pressure $P = P_B(T)$.

Index

www.ingramcontent.com/pod-product-compliance
Lightning Source LLC
Chambersburg PA
CBHW071453220526
45472CB00003B/788